the BIG and the SMALL

Volume II

Vignettes in Physics

A Series by G. Venkataraman

Title	**Abbreviations**
A Hot Story	*A Hot Story*
At the Speed of Light	*Relativity*
Bhabha and His Magnificent Obsessions	*Bhabha*
Bose and His Statistics	*Bose*
Chandrasekhar and His Limits	*Chandra*
Quantum Revolution (3 vols.)	
1. *The Breakthrough*	*QR I*
2. *QED: The Jewel of Physics*	*QR II*
3. *What is Reality?*	*QR III*
Raman and His Effect	*Raman*
Saha and His Formula	*Saha*
The Many Phases of Matter	*Many Phases*
Why Are Things the Way They Are?	*WATT*
The Big and the Small (2 vols.)	
1. *Journey into the Microcosm : The Story of Elementary Particles*	*Volume 1*
2. *From the Microcosm to the Macrocosm : The Fascinating Link between Particle Physics and Cosmology*	

the BIG and the SMALL

Volume II

From the Microcosm to the Macrocosm

The Fascinating Link between Particle Physics and Cosmology

G. Venkataraman

Universities Press

The Big and the Small (Volume – II)

Universities Press (India) Private Limited

Registered Office
3-6-747/1/A & 3-6-754/1, Himayatnagar, Hyderabad 500 029 Telangana, India
E-mail: info@universitiespress.com

Distributed by
Orient Blackswan Private Limited

Registered Office
3-6-752, Himayatnagar, Hyderabad 500 029 Telangana, India

Other Offices
Bengaluru, Bhopal, Chennai, Guwahati, Hyderabad, Jaipur, Kolkata,
Lucknow, Mumbai, New Delhi, Noida, Patna, Vijayawada

First Published 2007
Reprinted 2014, 2018

ISBN: 978 81 7371 574 7

Set in Caxton 10/12 by
OSDATA, Hyderabad 500 029

Printed at
Graphica Printers & Binders
Hyderabad 500 013

Published by
Universities Press (India) Private Limited
3-6-747/1/A & 3-6-754/1, Himayatnagar,
Hyderabad 500 029 Telangana, India.

CONTENTS

PREFACE

To the young reader

This book is the second part of the two-volume set entitled *The Big and the Small*. The first volume, *Journey into the Microcosm: The Story of Elementary Particles*, dealt essentially with the *small*, i.e., elementary particles. The present volume first takes the reader to the other extreme, the *big* or the cosmos, and then wraps up the story by briefly describing the fascinating connecting link between the two. In a sense, the *small* and the *big* are like the two strands of the DNA; in this case they form the strands of the DNA of the cosmos itself. And without having some idea of the two strands separately, the interconnection just cannot be understood.

In the earlier volumes in this series, we had glimpses of bits and pieces of the wonderful tapestry called the cosmos. We now stand back to take a parting look at the entire picture, its towering majesty, incomparable grandeur and indescribable beauty—an unfathomable mystery that continues to keep scientists engaged in the quest for the secrets unrevealed by nature.

I hope you the reader get as much enjoyment reading the book as I did in writing it. All the best!

To the adult reader

All my previous books in this series began with a standard message that has remained constant thus far. This time, I wish to say something different. The reason is simple —this is the last book in a series that I have been immersed in for over a decade, and I would therefore like to look back before saying goodbye.

As I have explained in my earlier books, I started this series in memory of my dear son Suresh who left this world rather unexpectedly at an early age when he was so full of promise. Since then, I have tried to wander in the world of physics, talking to Suresh as I went along. On the face of it, this might seem like silly make-believe; but then there were immense rewards for living in this artificial world of my own creation. Many young people in different parts of the country wrote to me, responding, I guess, the way Suresh would have, some with doubts, some with questions and a few seeking advice. This has happened in a sustained manner throughout, and the letters have come from all over the country.

I remember one case particularly because it was one where I had a direct contact with the correspondent. The incident happened about six years ago when my dear wife Saraswati was terminally ill. It was around 8 p.m. and I was busy attending on her—those were difficult days indeed. Suddenly the door bell rang. I was irritated but had no choice except to open the door and see who it was. Standing on the door-step was a young student, about seventeen years old. The boy identified himself and started plying me with intricate questions about the entropy of black holes! Imagine that!! Here I was nursing a patient who did not have long to live, and there was this young man wanting to know all about black holes—nothing could have been farther from my mind at that moment. Yet I paused for a second to reflect on the impact my running 'dialogue' with Suresh was having. Over the years I learnt, with much satisfaction of course, that many young students actually began to like physics after reading the VIP series and went on to do their PhD.

I started the series targeting specially the young, but as the years went by I discovered that I had more fans among the older people than among the young. This came as a total surprise to me, but it only showed that these so-called adults were young in spirit. There were other interesting experiences as well. Three years ago I was in Coimbatore airport waiting for a flight to Bangalore. A stranger walked up to me with a copy of my book on Homi Bhabha in his hand. Somehow he had learnt that I was the author and wanted to have a chat with me. I was amazed. He was not an isolated singularity; from time to time, I have received mail from people in different walks of life who had got attracted or who had been fascinated by physics after reading my books.

Among these I must mention one person in particular and that is Dr Abdul Kalam, currently our revered President. I have had the pleasure and the privilege of knowing Dr Kalam since 1984. In fact, after I moved over from the Department of Atomic Energy to the DRDO and was trying to set up a new lab named ANURAG in Hyderabad, Dr Kalam was my 'landlord', having given us some space within his sprawling laboratory complex. Dr Kalam was not only very supportive of all that I did while in ANURAG, but also a tower of strength on that dark day in May 1988, when Suresh suddenly left this world. Later when Dr Kalam moved to Delhi, he became my boss.

Dr Kalam had received his scientific baptism from Dr Sarabhai who succeeded Dr Bhabha as the Chairman of the Atomic Energy Commission. I knew Dr Sarabhai only slightly but had seen Bhabha in action from fairly close quarters. So it was inevitable that Dr Kalam and I would often compare notes about these two great sons of India. Dr Kalam paid generous compliments to the VIP series, and not surprisingly the book on Bhabha was his favourite. It was impossible not to be infected by Dr Kalam's perennial enthusiasm, and in their own way his deep sense of patriotism as well as his passion for making things happen in India have strongly influenced my thoughts as I worked my way through the many books.

While expressing my deepest gratitude to Dr Kalam for his sustained support, his companionship, and the inspiration he filled me with, I would also like to raise my hat to him for the extraordinary role he is currently playing as our President. What dynamism, vitality and extra dimensions he has injected into what is otherwise regarded as a decorative and figure-head post! When his name was being proposed, many scoffed at what they claimed was his lack of experience in politics. And how Dr Kalam has quietly shamed them all! I am sure that millions like me would say that he is one of the best presidents we have had, and that the country is indeed lucky to have him. What a tremendous job he has done, and continues to do, single-handed, in inspiring the youth of this country! Is there anyone in any walk of life, school teacher excepted, who sees about 300 to 400 school children practically every single day, even while travelling, and who inspires them the way he does with his special magic?

I have myself seen how Dr Kalam moves with young people, relates to them, makes them feel completely at home, inspires them to dream, and fills them with deep love for their country. If twenty years or so from now we have many stars in different fields of human endeavour including science and technology, we would have to thank Dr Kalam for it. I know Professor Raman used to love meeting students and talking shop with them, but I don't think he came anywhere close to Kalam's record.

Why am I bringing President Kalam into a series on physics? For a very simple and practical reason—he and I, in our own particular ways, are trying to inspire young people, or to ignite young minds as Dr Kalam might prefer to put it. Our methods may be different but we have the same objective: young people should take up challenges in science, technology and nation-building, with spirit, enthusiasm, courage and character. Spirit, enthusiasm, courage and character—it is these qualities in the ultimate analysis that lead to *real* achievements, both big and small.

Moving on to something different, the period of writing these books also coincided with my transition from active professional service to a so-called retired life . Active scientists abhor retirement and generally manage to have in place a 'retirement plan'. Luckily for me, destiny steered me away from such illusory paths. I felt that it was pointless trying to be as active in research as 25-year olds; one just cannot do that in science, at least in this country. It is better to accept gracefully one's 'senior citizen' status. However, that does not mean one should start leading a vegetable-like existence. And it is here that I must express my deepest gratitude to Sri Sathya Sai Baba for literally transforming my life in more ways than one.

In this series I have often quoted Weisskopf who said that knowledge without compassion is inhuman while compassion without knowledge is ineffective. Compassion—that is the key word. When Baba entered my life, I realised that the qualities of the heart are even more important for a human being than worldly intelligence or scientific prowess.

Thanks to Baba's kindness and generosity, I was accepted in the Sri Sathya Sai Institute of Higher Learning as an honorary staff member, which I have been ever since my superannuation in 1992. In the beginning I used to teach physics-related subjects but in the last few years I have been teaching a course listed in the curriculum as 'Awareness'.

Baba often tells his students that education is not to be imparted merely to enable students to eke out a living; rather, education is something that ought to prepare them for life, a life to be lived as a decent, compassionate and responsible human being. As Baba puts it crisply, education is not for living but for life; and life must be lived so that others are served and society benefited. The *awareness course* that I have been involved with stresses this point in various ways, and in this sense is quite unique. I honestly believe that its equivalent is not to be found in any other university either in India or abroad. In saying this, I do not mean to boast but to simply state a fact.

Let me explain. In this course we try to help the student to (a) see everything in a larger perspective, and (b) relate meaningfully to society, Nature, and God. This is far from a wasteful or foolish exercise. Scientific knowledge has made man very powerful, and if science is misused it can lead to disastrous consequences. Today, many big drug companies in the West are quite reluctant to do research on drugs and vaccines that can save millions of lives in the poorer countries. Their argument? "We would not be able to recoup our expenditure." I will not go into a discussion of how shallow this argument is. I refer you to the BBC documentary archives—look out for a series entitled 'Pills, Patients and Profits'. You will then know what I am talking about.

A word now about the patenting business behind which many multinational drug companies hide. About twenty years ago, there was only what was called process patenting. This meant that if a drug company made a medicine that had a particular chemical formula, what was patented was the process by which the drug was made. In the eighties, many enterprising Indian pharmaceutical companies found alternate ways of making the same drugs and at lower cost too. This was a great boon for large parts of the world and soon the pharmaceutical giants in the West were facing deep trouble. Those affected used their political clout and got the patent law changed from process patent to product patent. What this means is that now a patented drug cannot be made by *any* process whatsoever. Only one company (or its franchised agent) can make the drug and reap all the profits. For the record and by way of balance too, I should also mention the heroic efforts of Dr Virginia Hale of America who is using her knowledge of pharmacology for public good. After years of service in a leading drug company, she quit to found her own. Her purpose: to make drugs needed in the Third World at as low a cost as possible so that relief could be brought to the suffering in that part of the world. Thus it is that she has preoccupied herself with a drug for *kala azaar*, a dreaded disease widely prevalent in Bihar and other parts of India. Dr Hale took this step when she saw for herself the plight of children in North India, afflicted by the disease.

The business of patenting has gone to ridiculous limits, as one can see from the efforts to patent turmeric, neem, etc. Incidentally, thirty or so years ago, if a crystallographer solved a crystal structure of say a protein and submitted a paper to *Acta Crystallographica*, the best peer journal in the field, the editor would summarily reject the manuscript if the atom coordinates were not presented in full. Today, all that has changed. The authors can withhold quantitative structural data on the grounds that it is intellectual property, not to be revealed. The data is then sold, jointly by the scientist/s and the institution concerned to drug companies for a tidy sum!

You know something? When Sir J. C. Bose founded the Bose Institute in Kolkata at the beginning of the twentieth century, he declared emphatically at the inauguration that no discovery made in his institute would ever be patented! Bose added that he was merely following a hallowed tradition of this country.

There is a reason why I am bringing all this up. As Bose pointed out, in our tradition, knowledge is never held back; instead it is given away. In basic science this is very important, particularly if a discovery is one that would benefit humanity. This precisely is where Weisskopf's remarks cited earlier become profoundly important.

What I am getting at is that unless the imparting of worldly knowledge is balanced and unless it is blended gently with some information about the problems faced today by society, there can be serious problems. People hardly realise that over 80% of today's crime is committed by the educated. Some are involved directly, while others aid and abet silently. I mean, take the computer virus. Is it the poor, the ignorant and the illiterate who produce it? Or, take violence. Today, violence is expressed largely through the use of bombs and guns. Do you know that a few months after the recent Iraq war, when the dust had not yet settled down—it still hasn't even as I write this—there was a big arms bazaar in London at which $800 million worth of sophisticated arms were sold? No one bothered about this except for some protesters. But later, if there is genocide in some part of the world, the very same countries that actively promote small-arms trade would cry buckets (or would at least appear to do so), and resolutions would be passed in the UN Security Council. No wonder an academic in the West lamented, "All that our universities now do is to produce clever devils!" This may sound like an overstatement, but it does highlight the fact that apart from stuffing the brain, universities today are, in the name of competition, trying to outdo each other in killing any trace of sensitivity in their students. But then, as Gandhiji said, in the midst of darkness light persists.

Cut back to the late sixties of the last century. The Vietnam war was raging fiercely and I was in America then for a brief spell. While some American professors were gleefully accepting military funds for research (that produced deadly weapons like the naplam bomb, claymore bomb, etc.), others chose to forgo lucrative grants so that their conscience would not be burdened. That is what character is all about and that really is what Weisskopf was talking about. It is *not* an irrelevant point, particularly in today's dog-eats-dog world.

Recently, just before Independence Day, I heard on the radio a panel discussion. The subject was "Are Gandhi and Nehru relevant today?" I was amazed to find that most panellists and also members of the audience (mostly young people) thought Gandhi was fine but his ideas were no longer relevant. It seemed that people thought of Gandhi mainly in terms of his advocacy of simple life, his strong views against overdependence on technology, etc. It seemed to have escaped the attention of almost everyone that it is the *basic values* that Gandhi stood for that are more relevant today than anything else.

Gandhi cautioned us against the following:

wealth without work;
pleasure without conscience;
knowledge without character;
commerce without morality;
business without ethics;
science without humanity;
religion without sacrifice;
and politics without principle.

Don't we see all of this very much in evidence today? There is, for instance, a close connection between crime and politics—how on earth did this come about? After the recent rash of corporate scandals in America, a business school professor asked: "Whatever happened to ethics and morality that we used to teach as a part of our curriculum?" The answer is clear, is it not?

One of the towering characters in the epic *Mahabharatha* is Bhishma. We all know about him, about his infinite wisdom and other awe-inspiring qualities. But we also know that at a crucial moment, he chose, for his own reasons, to remain silent when he should have spoken out loud and clear, especially in view of the status and respect he commanded. Scholars spend a great amount of time criticising him. But how many of them point out that Bhishma was really a case study for modern intellectuals to learn from? I can say with first hand knowledge that if our university system is in shambles today, it is entirely because our intellectuals in the post-Independence period chose, for their own selfish reasons of course, to remain silent when the entire system was being mindlessly ravaged. Late Professor Sampath, a distinguished gentleman and a wonderful human being (he was Director IIT Kanpur, Member UPSC, President Computer Society of India, Member Electronics Commission and, some years before me, Vice Chancellor of the Sri Sathya Sai Institute of Higher Learning), once showed me a book which was a compilation of articles by Vice Chancellors of our various universities. Ninety percent of them dripped syrup in praise of the establishment, pure hypocrisy of course. Professor Sampath remarked, "Without sycophancy, one cannot climb higher, to the post of a Governor, for example. Only three Vice Chancellors dared to call a spade a spade, and they were promptly sidelined."

Take a simple thing like ragging. Everyone knows what happens and how it provokes extremes of uncivilized behaviour in senior students. In fact, so serious is the matter that it was once even banned. But since then slackness has set in, and we are stuck with an unwanted and barbaric ritual that has no roots whatsoever in our tradition. I just ask all the advocates of this 'innocent' initiation ceremony: Would Gandhi have approved of it?

It is not as if intellectuals alone are silent. Take our media. One leading metropolitan newspaper makes me wonder whether it is being published from India or some rich country. I mean there is hardly anything about the *real* India. If you want to know what *real* India is, try and get hold of a book that was published in the early 1940s—the name of that book is *Our India*, and the author is Minoo Masani. Sixty years have passed since Minoo Masani wrote the book.

Yes, we have become independent, and the statistics have changed too. The population in Masani's undivided India was 40 crores; today in divided India it is more than 100 crores. But the gap between the haves and the have-nots remains very much as before. Basically it is because of a sharp rise in the lack of sensitivity.

I recall a scene in (I think) 1983. I was in the auditorium of the Indian National Science Academy, Delhi, and the conference on Excellence in Electronics (with which I was actively involved) was taking place. The session on Electronics and Rural Development was in progress. One learned professor of computer science was waxing eloquent about how a doctor in a village could wire up a patient, how all the signals from the patient could be instantly transmitted via satellite to Jaslok Hospital in Bombay (then the best-known super-speciality hospital in the country), how the doctors there would diagnose what the problem was and flash back the information (via satellite of course), how the rural doctor could then treat the patient, and so on. I was laughing silently. I was then living in Kalpakkam (working in the Indira Gandhi Centre for Atomic Research) which is sandwiched between two villages—Sadras founded by the Dutch way back in the seventeenth century, and Pudupattinam, a fishing village. I knew exactly what villages were right then, in 1983 that is. It was clear that this learned professor had not visited a village even in his dreams. Just then Professor Sampath got up, adjusted his spectacles as he usually did, and said, "Mr Chairman, Professor N— has made a wonderful suggestion. It shows how powerful electronics is and how it can potentially solve many knotty problems. But, Mr Chairman, I personally believe that where rural India is concerned, it would be far simpler to teach our villagers to be healthy and give them all the assistance required to remain so!" There was a thunderous applause.

Gandhiji used to say that when a rupee, especially of public money, is spent one must, before spending that rupee, call to mind the face of the poorest person one knows and ask: "In what way will this expenditure benefit the life of this poor man?" Do we ever do that these days? Look at the way public money is being squandered all around us. As Baba often says, if we are a poor country, why were we repeatedly invaded? No, we have so much of so many things, including the incomparable blessing of a large population. We really do not need to borrow so heavily from the World Bank, IMF or other such agencies. A lot of the money needed could be internally generated by self-discipline. But who cares?

Here I am reminded of a conversation with late Mr V. K. Narasimhan who was a giant in the field of journalism. He had been associated with most of the leading newspapers of the country, and during the Emergency wrote fiery editorials unmindful of the consequences. He told me that when he was in Delhi, he once had a conversation with a VIP who was then a big shot in the Planning Commission. Mr Narasimhan knew the VIP very well. The conversation began with a comment from Mr Narasimhan to the Planning Commission mandarin:

"You know, you people in the Planning Commission have got it all wrong. You are not doing the planning properly."

"What do you mean we are not doing a proper job? Are you not aware of the tremendous amount of research that goes on before we draft plans?"

"All that is of no use; that research data does not convey anything about the *real* problems of the country."

"Come on, you must be joking."

"No I am not."

"Then what do you suggest we must do?"

"If you *really* want to plan for India, then all of you planners must get out of your air-conditioned offices in Delhi, go to some remote village in Rajasthan, in summer, sit in a hut there, on the floor, and then start your planning business."

"We can't do that! We would all be dead!"

Sharp-witted that he was, Mr Narasimhan came back with a wonderful repartee: "It would be no great loss to the nation!"

Humour apart, there is much truth in what Mr Narasimhan said, I mean about realistic approaches to society's problems. Such a need exists in the scientific field also. Just look at what happened recently when a new director had to be appointed for the TIFR. A search committee was appointed and the committee said (effectively): "We looked up and down, everywhere, and we have no option but to recommend the appointment of ... who is a distinguished NRI scientist."

Personally I have nothing against this appointee or even NRI scientists—I have many good friends among them. But here we are talking of the TIFR. Read the book on Bhabha in this series. What do we find there? Bhabha started the institute so that future leaders of science would emerge from there. Indeed they have—you do not have to take my word for it; just look at the output. At least four of its scientists have been elected Fellows of the Royal Society, which is saying a lot. I will not make a song and dance of this issue because it has been much debated in the Indian science community and in the pages of the journal *Current Science.* But this I will say: the distinguished members of the search committee claimed that a suitable candidate could not be found within India. This is no doubt a disputable statement but let us, for the moment, grant that it was indeed the sad truth. Then it means that the people who took charge of the destiny of science in this country after Bhabha left the scene did not quite deliver the goods. If Bhabha could achieve so much when he had such meagre resources, how could there be a slide when more support and resources were available? Interestingly, it turns out that the members of the committee were themselves actively involved in shaping the future of science in the country in the post-Bhabha period. That says a lot, does it not?

There is another thing I am bitterly disappointed about. Somehow, where grooming of talent is concerned, most of our preoccupation has been with metropolitan/urban areas and with the children of the better-offs. This is a huge country, and a substantial portion of *our* population lives in non-metropolitan areas. That really ought to be our 'catchment area' but that is not the way the mandarins of science seem to look at matters. I know from personal experience that first-generation scientists hailing from deprived and under-privileged families are quite comfortable staying back and do not feel the compelling urge to emigrate. That is an additional bonus. The present neglect of the hinterland is happening because *real* India is often out of the radar screen of the elite and that of the powerful who are busy with their own agendas. Sarabhai used to say that China and India would be the great powers of the 21st century because both of them would have huge reservoirs of youth. Having a substantial reservoir is no doubt necessary. But is that alone sufficient? Is it not also important to take steps to *ignite* the young minds suitably?

Since this series is on physics, I must pause here to raise my hat to another good friend of mine, Professor Richard Riesz, known to many of us simply as Dick. I first ran into Professor Riesz one hot afternoon on the streets of Madurai, where I was on a brief visit. I think it was in 1963. Since then Dick and I have been quite close. I should mention here that prior to coming to India, Dick had been a staff member of the famous Holmdel Lab of the Bell Lab chain, referred to in this book. What I am trying to drive at here is that here was this man, busy doing frontier research in one of the best labs in the world, and one fine day, after hearing the call, decided to give it all up, travel across the world to dusty Madurai and teach physics to students hailing mostly from what I have referred to as the hinterland. Of course physics was no doubt already being taught in many of the colleges there, but Dick brought an entirely different flavour to it. I know this because I have personally seen how he patiently built up his post-graduate department, step by step. Dick spent nearly twenty-five years away from his homeland helping the cause of physics in India when he had really no reason to, except that he had heard the voice of his heart. Lovingly supported by his charming wife Joyce, Dick completely adapted himself to local conditions, relishing south Indian cuisine and Carnatic music, learning Tamil and a bit of mridangam too.

Twenty-five years of hard work in dry and dusty Madurai cannot go in vain, and I have seen for myself how many of his students coming from deprived homes, scheduled castes and tribes, and under-privileged families, flowered later in physics and applied physics. By the way, I must also mention that it was Richard Riesz who produced all those wonderful photos that recreated many of Raman's classic experiments, photos that I desperately needed for my biography on Raman, *Journey into Light*.

I might add that Dick was not preoccupied solely with teaching, research and building up his lab. He also did projects so that he could raise money. Many industries in the Madurai area benefited from his expertise. And the money he got was ploughed back not only into the physics department but also into social programmes. For example, he had a computer centre for training handicapped people in computer skills. Dick finally returned to America not because he was tired of this country but because of the compulsions of advancing years. Even so, he keeps coming back now and then, just to keep in touch.

Professor Riesz, India has not forgotten you! No doubt you will always live in the hearts of your students, but let me, on behalf of the Indian physics community, tip my hat to your wonderful contribution that came from your genuine love for physics and for India. Have a nice and relaxed life back home!

Okay, now why did I bring in Kalam, Richard Riesz and others? The point I am trying to make is that there are two things in life: responsibilities and rights. No doubt we have a right to all sorts of things but we also do have responsibilities. When I was young, all the elders I knew talked mainly of responsibilities which is not really surprising because it was a part of our tradition; indeed elsewhere too there was more stress on duties than on rights. Have we not heard how character development was stressed in well-known schools of England and also the famous saying "England expects every man to do his duty"?

I remember how, in the early days after Independence, Jawaharlal Nehru used to often say, "Freedom brings responsibility." Curiously, our Constitution as passed in 1950 mentioned only rights and carried no reference to responsibilities. In the early 1970s our Parliament tried to rectify this and a Committee drew up a list of what it felt should be the fundamental responsibilities of every citizen. When the idea was submitted to the Parliament and a constitutional amendment was proposed, all the politicians felt that responsibility should not be made mandatory but optional—they wanted an escape route!

The rather unsavoury situation we have at present is largely because of large-scale indifference to values and responsibilities. Many feel that the situation is too far gone and say, "Listen, I've got to survive. I don't care in the least about what you are saying." Seems reasonable except that one really cannot blame others if disaster finally envelopes all, including those who seek exemption from values and responsibilities. I don't know how you feel but it is my firm belief that if even 20% of our people staunchly adhere to values, nearly half of our problems would disappear without the expenditure of a single paisa. How much does it cost the public exchequer for people to say: "No, I will not smoke and drink, I will not throw litter, I will exercise regularly, I will not spend too much time watching TV, I will not eat junk food, I will be honest, I will work hard and not waste time, I will avoid conspicuous spending, I will not be a party to bride burning, I will not encourage corruption in any manner whatsoever, I will not indulge in female infanticide, I will not cheat", and so on? The government cannot do this; we as individuals have to do it and we can only when, as Shakespeare said, we are determined to remain true to our self. And that calls for a commitment to values, integrity and character.

Sadly, values seem to have vanished from almost all spheres of human endeavour and that is our biggest problem. We seem to feel that values are a matter of private concern and that it is for the individuals to follow them or not to do so. This is *not* true. Everyone has a responsibility in sustaining and nourishing values—parents, elders in the family, teachers, community leaders, administrators, politicians, judges, etc. No one is exempted.

I was once talking to the cricket legend Sunil Gavaskar. He agreed with me about the disappearance of manners and decorum on the cricket field, and along with it, a sense of fair play as well. He responded to my comment by observing, "Yes, we once had the phrase—it is not cricket!"

Human beings do not live by bread alone. This was the wisdom instilled into young people of this land thousands of years ago. Later, Christ reminded humankind once again of this truth. I recall my own teachers who were simple people, very much committed to their work. I don't quite remember what they taught, but I certainly learnt a lot by observing them and the way they lived. I remember still my father's elder brother, a school teacher in a small town in Ramnad district in south India. He died way back in 1948. Believe it or not there were a thousand people at his funeral—that is the respect he commanded in the community.

Having been inspired by Gandhi all along, I have always been, in a vague sort of way, supportive of values but it was only when I came to Sathya Sai Baba that

I had the occasion to learn what I really should have learnt long ago. He opened my eyes and I realised for the first time that we all are what we are because of society. People complain all the time about India, especially those who want to leave. But I ask: "Have you heard of countries in Africa where even now in the 21st century, there are no schools, no proper transport systems, no reliable means of communication, very few hospitals or hardly any universities? In many countries, people do not have even enough to eat. How much better off we are compared to them! How then can any right-thinking educated person say that our country has not done anything for us?"

Yes, Baba made me aware not only of my higher responsibilities but also of the nobler purpose of life, especially as enshrined in our tradition. Inevitably this took me to Aurobindo and his writings, and that helped me understand how one must blend ancient values and traditions with modernity (which, incidentally, is the primary objective of the Sri Sathya Sai Institute of Higher Learning founded by Baba). This blending is an important sociological matter in itself, but I shall not go into all that here. I shall content myself with saying that but for Sathya Sai Baba, I would not have found true fulfilment in life. I am immensely grateful to him for making me realise the higher and nobler purpose of human existence and enabling me to do something useful for society, at least in the last part of my life. The fourteen years that I have spent in Puttaparthi after superannuation have truly been the best years of my life. From Baba I have learnt that the best thing one can do in life is to *love all and serve all*. Offering to him my deepest gratitude, I say my final goodbye to you the adult reader, for your sustained and loving support throughout this series.

All the best and God bless you!

As I wind up *Vignettes in Physics*, I offer with much love the entire effort that went into this series to Baba, who brought true meaning to my life.

G. Venkataraman
May 2006

ACKNOWLEDGEMENTS

Every book published previously in this series carried an acknowledgement specific to that particular book. As I bring the curtain down on the entire series, it is only fitting that I place on record my deepest gratitude to all those who in various ways made it happen.

Right at the top of my list are the publishers, Universities Press, who have not only been most understanding, kind, courteous and considerate, but also extremely magnanimous in giving me the fullest support, right from the word go. What is more, they never made even the slightest effort to dictate, make demands, or influence my writing. I have read a lot about how authors (in the literary world) have had problems with agents, editors and so on, but in all these years I have not faced such a problem even once. I don't think it is all a matter of extraordinary luck; rather, I firmly believe it is due entirely to the publishers, and it is for this reason that I begin with them. Thank you Universities Press for your staunch support to Indian authors who write quality books on topics related to science and technology.

Even the Good Lord requires instruments for carrying out His Missions, and so it is no surprise that publishers need them too. Where I am concerned, Madhu Reddy has been that instrument all the way, as has been Rukshad Banaji in the early days. Madhu, you have been such a tower of strength throughout, including the exciting moments of the launch of the book on Bhabha by Homi Bhabha's brother. As for you Rukshad, I enjoyed all the sparring during which you taught me many things, including (very subtly of course!) that I had better not think too much of myself! By the way, the wonderful book containing numerous backroom stories of particle physics that you gave me as a gift when you left Hyderabad, has inspired this last volume on the Big and the Small no small extent.

Books, especially on science, cannot be written just off the cuff or in a casual manner; they need a lot of researching and background material. During the period when I worked on this series, it so happened that I did not have access to as good a library as I was used to when I was in BARC and the IGCAR, Kalpakkam. It was Ratnakar, then of the library of the Raman Research Institute, Bangalore, who helped me out of this difficulty. Those were days when one could not dream of e-mail, internet, web sites, etc. Indeed, even photocopying was quite a luxury. Despite all these handicaps, Ratnakar produced not only all that I was looking for but also much more that I was not even aware of. Even more valuable to me were the hours we spent together, talking about science and vanishing values. In recent times, Girija Srinivasan, also of the RRI Library, gave me the help that Ratnakar earlier used to, in particular through her skilful surfing on the web. Thank you both and may God be with you always.

And now to the academics. Many have acted as lighthouses for me in this long voyage but on this occasion I need to particularly remember two of them. The first is my old colleague and constant friend, Professor V. Balakrishnan of IIT Madras. I learnt a lot about many things, especially about the art of communication from Professor Balakrishnan. Bala also influenced my thinking itself in many ways. Incidentally, I owe the series title Vignettes in Physics (VIP) entirely to him.

The other person to whom I owe a similar debt of gratitude is Professor G.Srinivasan of the Raman Research Institute, who, incidentally, happens to be my younger brother! The curious thing is that Professor Srinivasan and I hardly ever discussed either any particular book or even the series as a whole. And yet, whenever I wrote I always felt: 'Little brother is watching you!' I do not mean this in the Orwellian sense—far from it. What I mean is that I always had the feeling: "Would he approve? Would he like what I have written?" All this not only because he is a far, far better physicist than I and has vastly superior skills in making science come alive, but also because of his uncompromising sense of values and impeccable taste in all good things, and particularly in science. My greatest surprise came when I learnt that Professor Srinivasan had, unknown to me, recommended my books to many of his students! That, I should say, was something rather like Vashishta awarding the seal of approval to Vishwamitra!

I must also mention three individuals who occupy a special place in my life. First there is Suresh—he inspired this series. Then there is my late wife Dr Saraswati Venkataraman, herself a physicist. When Suresh left us and we were rudely and suddenly left all alone, I owed it to her to spend more time with her than I had in the past. But there was this call of the series. She understood and made sacrifices so that I could do what I had set out to do. Finally there is Bhagavan Sri Sathya Sai Baba to whom I have already made references elsewhere. Here I would like to just say: Thank you Baba for teaching me that the best way to love God is to love all and serve all. That one teaching has helped me immensely not only to complete this series against many odds, but more importantly to make something of the last part of my innings.

In conclusion, let me in the best traditions of this land, say:

Loka Samasthaa Sukhino Bhavantu.
Om, Shanti, Shanti, Shanti.

G. Venkataraman

1

PICKING UP THE THREADS

Welcome back! When we parted company last (that was at the end of Volume 1*), I was telling you about the great triumphs in the field of elementary particle physics. In my opinion, the three great achievements of physics in the twentieth century are: (1) Relativity, inclusive of both the Special and the General, (2) Quantum Mechanics and Quantum Field Theory, and (3) the Standard Model of elementary particles. This time, I shall try and describe how the physics of elementary particles would allow us to reconstruct the origin of the Universe itself and also describe its subsequent evolution!

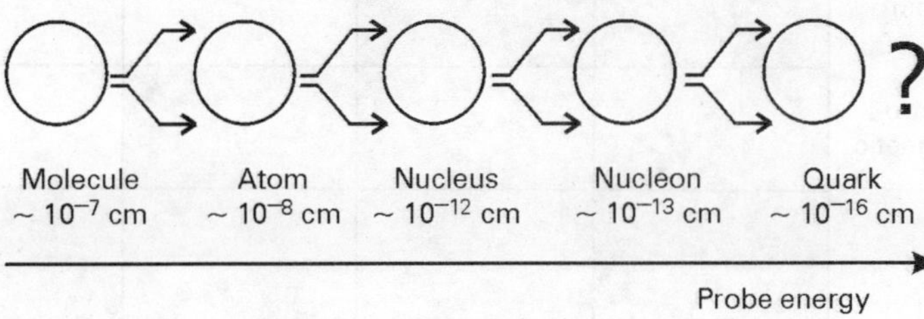

Fig. 1.1 *This figure illustrates schematically the basic trend of twentieth century physics, highlighting the fact that scientists were able to probe at smaller and smaller scales of length by going to higher and higher energies.*

Let me start with Fig. 1.1. This says that as one tries to probe matter on smaller and smaller scales of length, one has to use probes of higher and higher energies. During the second half of the twentieth century, physicists managed, using gigantic accelerators that reached up to several TeV of energy, to go down almost to 10^{-16} cm—all this I have described in considerable detail in Volume 1. At the end of it all it was realised towards 1985 or so that the physics of elementary particles, as then understood, could be compactly described via the so-called Standard Model (SM). I have already discussed the Standard Model in the previous volume. For your convenience its essence is captured again in Fig. 1.2.

* Volume 1 refers to the first part of this book entitled *The Big and the Small: Journey into the Microcosm.*

Generations	Quarks		Leptons	
First	u (Up)	d (Down)	e (Electron)	ν_e (Electron-type neutrino)
Second	c (Charmed)	s (Strange)	μ (Muon)	ν_μ (Muon-type neutrino)
Third	t (Top)	b (Bottom)	τ (Tau)	ν_τ (Tau-type neutrino)

Charges				
Electric	+2/3	−1/3	−1	0
Colour	Red YELLOW BLUE	Red YELLOW BLUE	Colourless	Colourless

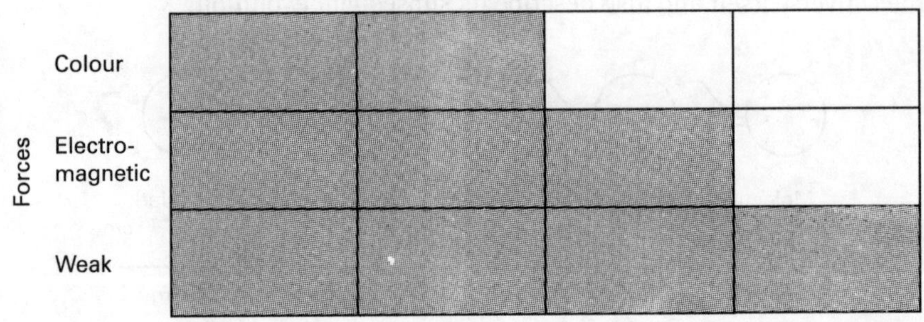

Fig. 1.2 *Summarised above is the essence of the Standard Model (SM) of particle physics in terms of three generations of quarks and leptons. The charges and colours (where appropriate) of these particles are shown, as also the forces they respond to.*

The Standard Model says that the basic building blocks of matter are six quarks and six leptons. All matter is supposed to be built up basically from these. The quarks and leptons come in three generations. The quarks have colour charge and fractional electric charge, while the leptons have no colour charge but only electric charge as indicated. The quarks basically interact via colour force (which gives rise to strong interactions), while the leptons interact via weak force and electromagnetic force (which in combination function as electro-weak force). Figure 1.3 describes the basic interactions in the language of Feynman diagrams. It is useful to recall these basics, even as we prepare to start on our new journey.

Two other important facts also must be borne in mind. First is that the forces mentioned above are mediated via bosons (in contrast, quarks and leptons are fermions). Secondly, parity is not conserved in weak interactions. As a result, the neutrino acquires a distinctive handedness, a concept that is illustrated in Fig.1.4. Of

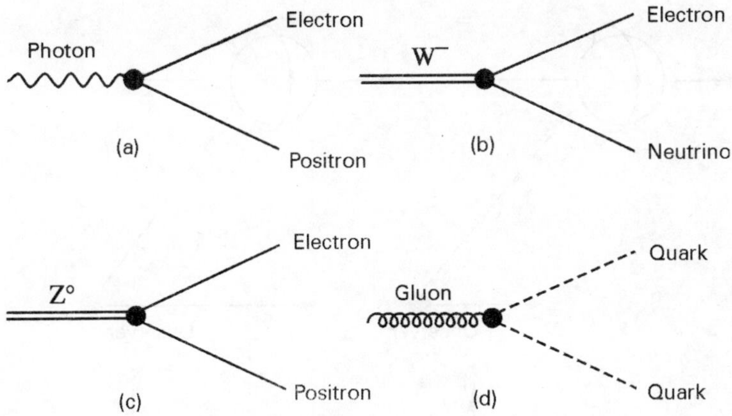

Fig. 1.3 *This figure illustrates the basic 'building blocks' of the Feynman diagrams for (a) electromagnetic, (b & c) weak and (d) strong interactions, each involving one vertex. The sample shown above is not exhaustive. More complex diagrams are essentially extensions of the basic diagrams involving many vertices.*

course, this conclusion about the distinctive handedness assumes that the neutrino has zero mass – this is what people believed at that time.

Don't be fooled by the simplistic description given above. It took a *tremendous* amount of effort by many brilliant minds, both on the theoretical and the experimental fronts, before something like the Standard Model could be constructed. The great feat of the Standard Model is that almost all of its predictions have been experimentally verified. One prediction that still awaits confirmation is the existence of the Higgs particle. No one doubts anymore the existence of this particle—it is just a matter of time before it is formally identified. One can almost bet that the Higgs particle would be detected as soon as the LHC or the Large Hadron Collider starts functioning in CERN, perhaps around 2007, may be a little later.

Thanks to the Standard Model, we are now sitting pretty with almost all known facts well explained. Does that mean that the SM is the end of the road? Far from it. First and foremost, you must remember that the model is only an ad hoc 'pasting together' of a theory of electro–weak interactions with a theory of strong interactions. The two are *not* meshed together in a natural fashion. Secondly, there are too many parameters in the SM. Thirdly, why should the quarks and the leptons form two distinct families? Could they not transform into each other? Fourthly, how are we sure that physics would remain the same when one goes from the SM regime, which is about 10^{14} eV to super-high energies such as 10^{28} eV? These are very valid points, and indeed a tremendous amount of work is going on right now towards answering these and similar questions. My job presently is to tell you something about this fascinating activity. In the process, we shall almost be taking a peek at the birth of our Universe! We would discover that to describe the Baby Universe we need what might be called the Ultimate Theory, the theory of physics that stands all by itself and does not rest on any other pedestal; that is to say, this Ultimate Theory would not require any other theory to explain

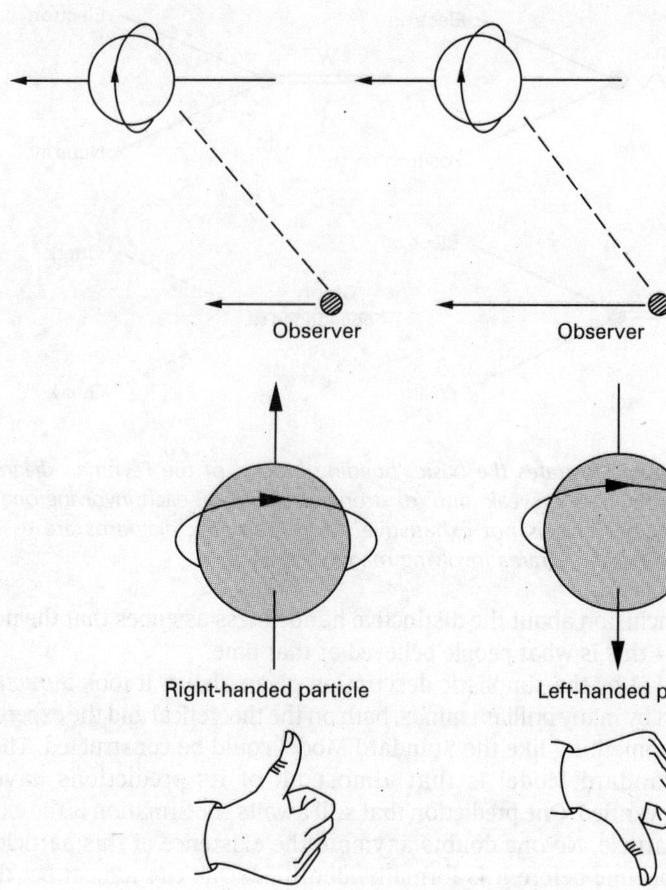

Fig. 1.4 *Illustrated above is chirality or the concept of handedness, which is important in the context of parity violation in weak interactions. First see the top. If a moving observer is overtaken by a faster moving electron, it appears to move away and the electron obeys the right-hand rule. What it means is that when the fingers of the right hand curl in the same direction as the spin, the thumb gives the direction of motion for a right-handed particle (see now the bottom part). If the observer overtakes the electron, it is a different story, and the handedness of the electron changes because the electron now appears to be approaching rather than receding. This handedness is illustrated at the bottom where the fingers of the left hand curl in the same direction as the spin; the thumb now gives the direction of motion of the left-handed particle. Chirality is important for particles that are almost massless.*

either its features or its constants. By comparison, the SM is what is usually called an *Effective Theory*. An effective theory works in its own domain but does not work when one goes to a deeper level. To give an example, where atomic physics is concerned, we can regard the nucleus as a point with charge *Ze* and mass *A*, but when we want to describe the physics of the nucleus itself, we cannot any longer treat the nucleus as a point—a new level of physics is required. Thus, atomic physics is an effective theory that rests on nuclear physics. Likewise, conventional nuclear physics is also an effective theory; it rests on the SM. And SM in turn is an effective theory that rests on the yet-to-be-discovered Ultimate Theory.

Please note that we go along we would often be using units that are convenient and relevant in the context of early Cosmology. Therefore take a good look at BOX 1.1 and keep those facts in mind.

BOX 1.1

When it comes to the Baby Universe, the Universe itself provides its own units of length, time, and mass. These are obtained by combining the Planck's constant $\hbar$, the velocity of light c, and Newton's gravitational constant G. These units called Planck units are given by:

Planck length $l_p = (G\hbar/c^3)^{1/2} \sim 10^{-33}$ cm

Planck time $t_p = (G\hbar/c^5)^{1/2} \sim 10^{-43}$ s

Planck mass $m_p = (c\hbar/G)^{1/2} \sim 5 \times 10^{-5}$ gm $\sim 10^{19}$ GeV

2

IT TAKES GUTS TO GO FURTHER

This chapter takes the first peep beyond the Standard Model, giving a glimpse of what are called Grand Unified Theories or GUT for short. The most exciting thing that happens when one graduates from SM to GUT is that one finds that the proton is no longer stable; it can decay but with a very long half-life, which is comforting because it is this long half-life that has made evolution of life possible!

Imagine a voyage of discovery. After sailing for a long time across unchartered waters, the explorers arrive at an island. They are charmed by the island – everything seems nice and tidy. However, the explorers soon realise that some distance away from the island lies a huge, unknown and fascinating continent. Forming into small parties, the explorers head for the continent. These parties land at different spots on the continent and start exploring the region around. Each group finds something amazing. From time to time the exploring parties meet and exchange notes, each group describing the new continent as seen by it. After comparing notes, the various exploring teams realise that they are rather like the proverbial five blind men who tried to describe the elephant by feeling only parts of it. However, unlike the five blind men, these explorers know very well that the continent they are exploring is not merely what they are currently seeing but something much more. This allegory more or less sums up the situation in particle physics today!

The Standard Model (SM), as we saw earlier, does a terrific job as far as it goes but is clearly not the whole story. In the next couple of chapters, I shall try and tell you about some of the efforts that have been made to go *beyond* the SM. Let me start with a class of theories that bear the generic name *Grand Unified Theory* or GUT for short. Recall from Volume 1 that SM is described in the gauge group language as

$$f = SU(2) \times U(1) \times SU(3).$$

As I told you earlier, soon after the SM came into existence people began to wonder: "Why should there be two separate theories, one for leptons and one for quarks? Why two types of building blocks? Why two types of forces, the electro–weak and the strong force? Why not just one type of building block and just one interaction force?"

Now what made people think like this? To appreciate this let us go back a bit in history. Initially, people did not know anything about the weak force; they knew only about the electromagnetic force. Then came Fermi who introduced the weak force by analogy with the electromagnetic force. Even so, the two forces, the weak and the electromagnetic, seemed so different. But with the work of Salam, Weinberg and Glashow, it became clear that though the electromagnetic and the weak forces appeared superficially different, they were actually two aspects of a common electro-weak force. This commonality lies hidden at low energies but clearly reveals itself at sufficiently high energies as Carlo Rubbia demonstrated most convincingly—all this I have already discussed in Volume 1. Guided by all this, people now asked, "Why can't the strong and the electro-weak forces similarly merge into one grand unified force at sufficiently high energies?" See Fig. 2.1. Seeing no reason why such grand unification could not occur, many went ahead to try and discover a GUT. In simple terms, all these people were looking for a gauge group *G*, the interaction associated with which comprehensively describes in one breath as it were both the strong and the electro-weak interactions. In this language the product group *f* describing the SM would be a subgroup of *G*.

Has such a group *G* been found? Unfortunately no, although people have been searching for over two decades. Many candidates have been proposed and for a while some of these even looked like being hot favourites; but to date there is no clear winner.

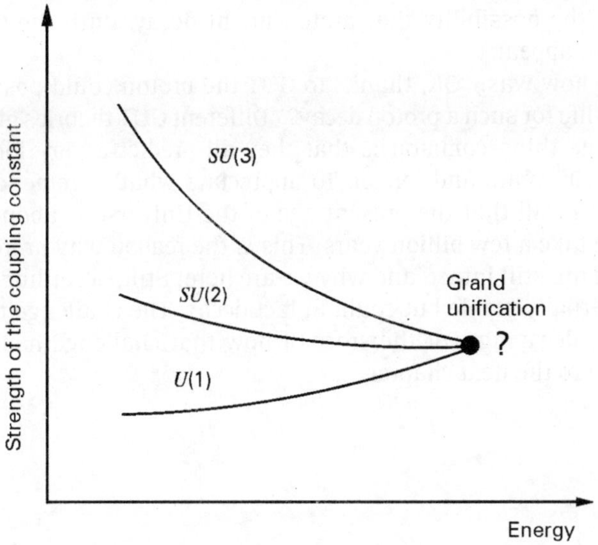

Fig. 2.1 *Shown schematically above is the anticipated energy variation of the three coupling constants associated with the three groups that provide the basis for SM. The hope and belief is that at some energy, perhaps very high and beyond the reach of today's accelerators, there would be a convergence of the coupling constants, indicative of grand unification.*

The details of the GUT-type of theories are quite complicated and we need not be bothered about them here. But there is one important fact I must call attention to: many versions of these theories demand that the quarks and leptons must not remain as independent families but must transform from one species to another. The interaction associated with this inter-conversion of particle species is supposed to be facilitated by a gauge particle, usually called the *X particle*. One crucial fallout of this feature is that the proton, which till recently was believed to be completely stable (like the electron), can undergo radioactive decay.

This prediction of the GUT class of theories excited physicists very much. From the thirties people knew that the neutron could spontaneously decay. That however was no great surprise. The mass of the neutron is much more than the combined mass of the proton and the electron; therefore the decay of the neutron is not forbidden on considerations of mass. The proton, on the other hand, could not decay in a similar manner into a neutron, a positron and a neutrino. The lower mass of the proton ruled this out. But what about a decay like say,

$$p \rightarrow \mu^+ + \nu_\mu + \text{energy},$$

or something like that? Such a decay could not be ruled out on considerations of mass alone. However, nothing like this or any other similar type of decay has ever been observed. People therefore assumed that the proton never decays, and then invented an argument to explain why the proton does not decay. After this, everyone forgot the possibility that proton might decay, until the GUT-type of theories began to appear.

The question now was: "OK, thanks to GUT the proton could possibly decay. What is the half-life for such a proton decay?" Different GUT theories give different estimates but one thing common is that they all predict a very long lifetime, something like 10^{28} years and longer. To appreciate what a tremendously long time this is, just recall that the present age of the Universe is about 15 billion years or so, give take a few billion years. This is the reason why protons created in the beginning are still intact, and why we are here! Still, according to GUT the proton is not eternally stable but could in fact decay. The challenge then was to catch the proton decaying. For the story of how that challenge has been dealt with, please turn to the next chapter.

3

WAITING FOR SOMETHING TO HAPPEN

This chapter offers a brief review of the major experiments designed to detect proton decay.

In the last chapter I told you that till the third quarter of the twentieth century, physicists did not ever entertain any doubt that the proton was stable; there was no evidence to the contrary. If protons decayed, then the Universe would be losing matter all the time, at least to some extent. There is, however, no evidence to show that the Earth is losing weight. All things taken together, it seemed (until recently) as if the proton was very much stable, just like the electron.

If the proton is stable, then there must be a law that says so. Many people believed that this was the case; they then surmised that Nature has in fact ordained such a law and named it *Baryon Number Conservation* – see also BOX. 3.1. We have seen in Volume 1 that conservation laws have a deeper basis. So, what was the basis for Baryon Number Conservation? No one could say what the basis was. For electric charge conservation, for example, global gauge invariance is the deeper underlying principle—I referred to this in Volume 1, Table 3.3. Does there exist a corresponding gauge principle for Baryon Number Conservation, and if so what is it? No one knew. There the matter stood until the seventies of the twentieth century. Physicists were too busy exploring a number of other interesting leads, and since the proton seemed to be quite stable they took it for granted that Baryon Number Conservation was true, whatever be the basis.

When people started investigating GUT-type of theories, the question of proton stability assumed great significance. Salam and Pati were among the first to try for a GUT, and in their version quarks could convert into leptons. In turn this implied that the proton must decay over a period of time, even if it be rather slowly. Salam was so excited by this discovery that he went around encouraging experimenters to look for proton decay. When he came to India he persuaded the Tata Institute of Fundamental Research (TIFR) to set up such an experiment. At that time TIFR was running many cosmic ray experiments in the Kolar gold mine pit (see ***Bhabha***), and a deep mine was the ideal spot to conduct the experiment suggested by Salam. So TIFR very quickly implemented Salam's suggestion. In the beginning not many took Salam seriously. But in due course when most of the different versions of GUT that were being proposed all indicated that the proton was not stable, looking for proton decay became a serious matter.

At this point I should point out that in principle, there could be three types of grand unified theories, as follows:

1. No baryon number is defined and the proton is therefore unstable.
2. There *is* a baryon number, hence also a corresponding symmetry, but it is broken and therefore the proton could decay. However, the decay time depends on the details of the theory.
3. There is once again a baryon number and therefore also a symmetry; however, the symmetry is not broken. Therefore, the proton is absolutely stable.

Keep all this in mind while you proceed. Figure 3.1 illustrates some of the different ways in which the proton can decay.

Let us now turn to the methods for the detection of proton decay. There are basically two methods. In the first one uses the fact that protons are present

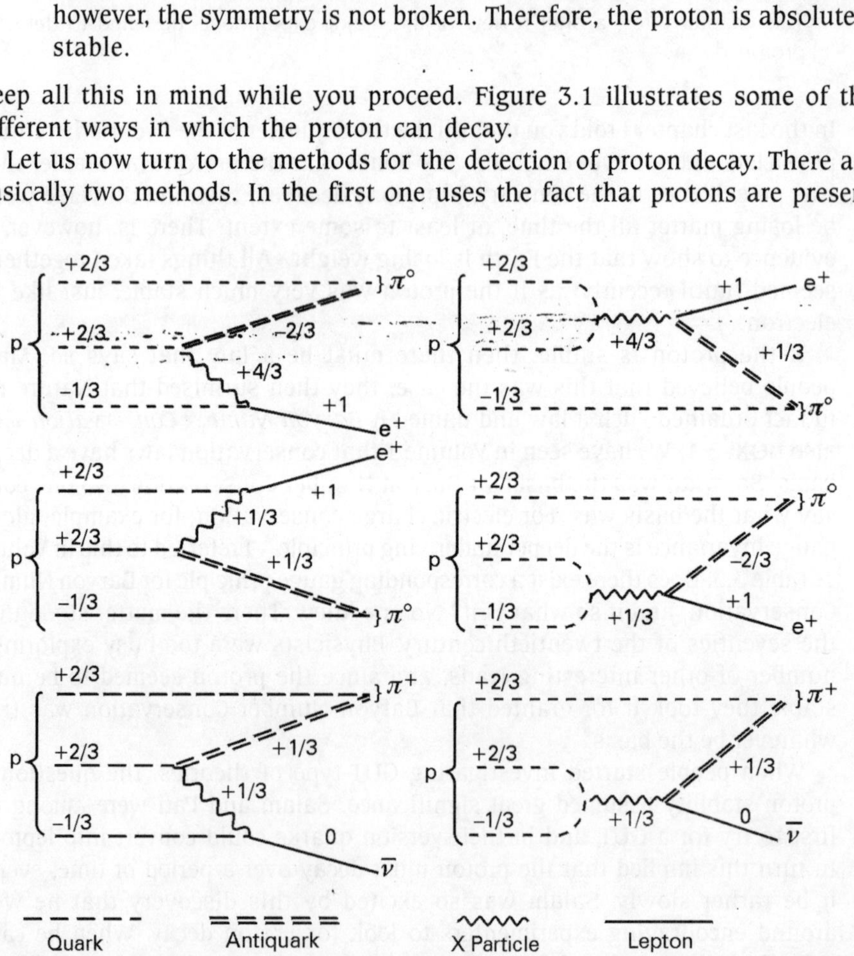

Fig. 3.1 *This figure gives the details of some of the different ways in which a proton can decay into a pion and a lepton. This happens via the emission and subsequent absorption of an exotic particle X. In the above cases this exotic particle is one of two types: (a) charge + and spin 4/3, or (b) charge + and spin 1/3. Many other such modes of decay are possible.*

in all nuclei. Hence, all that one has to do in principle is to keep looking at a very large number of nuclei and catch those that transform due to GUT physics. How does one catch such events? There are two ways of doing this. The first is by looking for the daughter nuclei that would result as a result of proton decay of the parent nucleus. This method is essentially indirect. The second method, on the other hand, is direct—here, one directly detects the particles emitted in the spontaneous decay of proton. All the experiments set up in recent years have invariably adopted the latter route.

In method 2 one basically has a very large volume of very pure water and 'keeps looking at it' with a massive bank of very sensitive photomultipliers. The latter can pick up tiny flashes of light produced by the decay particles. If the flashes *unambiguously* add up to the right signature, i.e., the signature associated with *genuine* proton decay, then presto, one has *actually* caught the proton decaying. It is all quite simple in principle but in practice there are many difficulties. In fact, detection of proton decay is one of the most difficult experiments ever performed. The first and foremost problem is that one must have a large quantity of water for observation. You might wonder why this is a problem—there is so much water everywhere; even in the desert there is water! It is not the quantity that is the problem but the ultra, ultra purity that is a must. As you well know, if one is not careful, microorganisms can easily enter water and flourish there. So, one must ensure that not only is the water extraordinarily pure to start with, but *remains so* during the *very long* periods of observation.

Next, there are the masking signals from background events. Cosmic rays constitute the principal source of worry; the problem can be minimised by going underground to a great depth or deep into the mountains—both have been done. Even after all this one can have problems, this time from the neutrinos!

We normally tend to think that neutrinos hardly interact with matter. Indeed, this is true (especially for low-energy neutrinos) and that is why the detection of the neutrino took such a long time. But when we are considering an experiment like proton decay, even this very feeble interaction capability of the neutrino can become a problem. You may wonder from what sources these trouble-making neutrinos can come. From many sources, such as the atmosphere, for example. Atmosphere? Yes, the atmosphere! How? That is simple. The cosmic rays coming from outside create many neutrinos in the upper atmosphere; this is happening *all the time.* These neutrinos produced in the atmosphere can penetrate easily even into deep mines, and although their interaction cross-section is small, over long periods of time many neutrino-induced events can take place inside the observation volume, sending confusing signals. Is there no way of getting rid of these atmospheric neutrinos? Yes, there is; one can set up the experiment on the Moon! Obviously that is not possible at the moment, and so for the present physicists have geared themselves to living with this neutrino background, devising criteria to distinguish genuine proton decay events from neutrino-induced events.

To give you a feel for the neutrino problem, let me mention the following: Let us say the proton decay half-life is 10^{33} years. As already mentioned, this is an incredibly long time. Suppose that the quantity of water in the experiment is 3000 tons. Then during one year of observation one would be lucky to detect *one* genuine proton decay event! The neutrinos on the other hand, can generate signals *once every three days or so*! So you see how tough the problem is! By the way, this 'neutrino difficulty' has recently proved to be a blessing in disguise. More about that later.

So what's the latest on proton decay? Well, experiments have been going on for many years, two of the most prominent ones being the IMB or the *Irvine–Michigan–Brookhaven* collaboration in America and the *Super Kamiokande* experiment in Japan. Both employ water detector, and to date no clear-cut and unambiguous event demonstrating proton decay has been found. What it means is that the proton either does not decay at all or its lifetime is longer than anticipated. Bit disappointing, isn't it? Yes that certainly is true where proton decay is concerned, but patience always is rewarded. This reward came on 23rd February, 1987, in a most unexpected manner. That day, both detectors observed a burst of neutrinos coming from a supernova explosion. This has led to some good neutrino physics! You have to wait till a later chapter for that story.

BOX 3.1

Proton stability was first formulated in 1929 by Herman Weyl who said: "It is plausible to anticipate that of the two pairs of components of the Dirac quantity, one belongs to the electron and the other to the proton. Further, two conservation laws of electricity will have to appear which state that the number of electrons as well as of protons remain constant." You might recall my mentioning Weyl in Volume 1 [see BOX 3.1, section 5.13 and BOX 10.1]. Weyl's proposal did not attract much attention at that time. In 1938, E.C.G Stuckelberg reformulated the conservation law as: "For all observed transformations of matter, no transformation of heavy particles (neutrons and protons) into light particles (electron and neutrino) have yet been observed. We therefore wish to postulate a conservation law of 'heavy charge'."Today, we call the 'heavy charge' as baryon number. In 1949 Wigner rediscovered this law and said: "It is conceivable, for instance, that a conservation law for the number of heavy particles (protons and neutrons) is responsible for the stability of protons in the same way as the conservation law for charges is responsible for the stability of the electron. Without the conservation in question, the proton could disintegrate just as the electron could (were it not for the charge conservation law), into a light quantum and a neutrino."

Both Weyl and Wigner, experts as they were in group theory, wondered whether there could be a foundation for this baryon conservation law based on symmetry. However, they did not explore this possibility and there matters stood till the GUT theories brought the question of proton stability to the forefront.

Recall from Volume 1 that the baryon number for the particle and the anti-particle have opposite signs. Thus, baryon conservation does not mean that the proton is immortal under all circumstances. If, for example, a proton comes into contact with an anti-proton, it can certainly annihilate. The conservation law merely prohibits the *spontaneous* decay of the proton into lighter particles. This is highlighted in some of the decays discussed below. Here, mass is given in units of MeV and charge is given in units of the charge of the proton.

$\pi^{o} \longrightarrow \gamma + \gamma$
Mass: $135 \longrightarrow 0 + 0 +$ Energy
Charge: $0 \longrightarrow 0 + 0$
Baryon number: $0 \longrightarrow 0 + 0$

$n \longrightarrow p + e^{-} + \bar{\nu}$
Mass: $939.6 \longrightarrow 938.3 + 0.511 +$ Energy
Charge: $0 \longrightarrow 1 + (-1) + 0$
Baryon number: $+1 \longrightarrow +1 + 0 + 0$

$\Lambda^{0} \longrightarrow p + \pi^{-}$
Mass: $1115.6 \longrightarrow 938.3 + 139.6 +$ Energy
Charge: $0 \longrightarrow = +1 + (-1)$
Baryon number: $+1 \longrightarrow +1 + 0$

In all the above cases, the decay conserves baryon number. Consider now a decay like $p \longrightarrow \mu^{+} + \nu_{\mu}$ mentioned earlier. In this case the baryon number on the left is 1 while on the right it is zero. Hence, by the law of conservation of baryon number this decay should be forbidden. This is what people believed till GUT came along.

4

SUPER PHYSICS

The aim of this chapter is to give a glimpse of the territory that lies beyond GUT. Many extensions have been attempted, and of these we focus here on three: Supersymmetry, Quantum Gravity, and Supergravity. This forms a first introduction as it were; these topics would crop again later in various places.

4.1 Introduction

The exotic title of this chapter is intended to call attention to some daring attempts to go even beyond GUT. Presently, I would like to discuss briefly the following:

1) *supersymmetry* or SUSY for short,
2) *quantum gravity*, and
3) *supergravity*.

Something even more radical is reserved for the next chapter. You will recall that the very first attempts at discovering GUT were based on finding a gauge group encompassing the group that describes the SM. I mentioned this in an earlier chapter. The physics discussed in this chapter seeks to go beyond all that. At best they form different bits and pieces of the final tapestry; at the other extreme they are valuable attempts to guide us into the future, though they may not be adequate in themselves.

4.2 Supersymmetry

The basic idea underlying *supersymmetry* (SUSY) is quite simple. It is that for every known building block of matter and for every known boson that mediates the forces between these elementary building blocks, there is a *superpartner*. This superpartner has the same attributes as the original particle, except for spin and mass. Presently I shall explain the concept in slightly greater detail, but first I must draw your attention to one important point.

In physics one is constantly aiming at economy in the number of building blocks. For example, Gell Mann introduced the idea of quarks in order to explain all the known baryons and mesons as composites of a small number

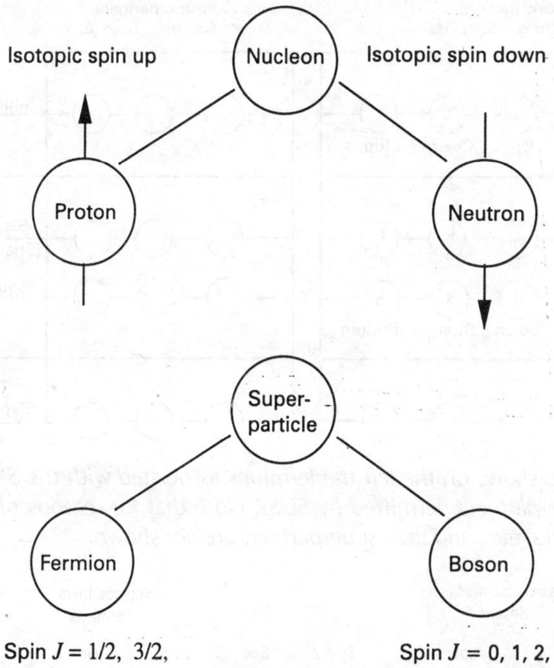

Fig. 4.1 *The purpose of this figure to highlight the fact that supersymmetry is a concept similar to isotopic spin. Just as, via the concept of isotopic spin, the proton and the neutron can be regarded as two isospin states of a nucleon, so also the fermion and its superpartner the boson can be regarded as two states of a superparticle.*

of particles. The SUSY idea obviously goes *against* that principle; in fact, it doubles the number of known particles. If in spite of such proliferation SUSY is considered important, then there must surely be some over-riding advantage; indeed there is. I shall come to that shortly, after explaining something about superpartners.

Let us start with Fig. 4.1. Here we have the proton and the neutron, and they are quite dissimilar. Yet, via the concept of isotopic spin one can regard these two particles as just two isospin states of a single particle, the nucleon. Turn now to the Standard Model. This calls for a set of fermions on the one hand, and a set of bosons on the other. They stand apart as two distinct families, and there is no way of looking upon fermions and bosons as being related, within the Standard Model that is. This is where SUSY helps. It helps in regarding the boson and its associated fermion as two states of the same superparticle.

Consider for example the graviton. You will recall that this is the quantum of the gravitational field. It is a boson with spin 2. Its superpartner would be a fermion with spin 3/2. A few words at this stage about the naming convention. The superpartner of a boson has a name with the suffix *no*. Thus, the partner of the graviton would be the *gravitino,* the partner of the photon would be *photino,* the partner of the gluon would be *gluino*, and so on. In the case of fermions, the naming convention calls for the addition of the letter *s* in front; thus the superpartner of an electron is a *selectron*, that of a quark is *squark*, that of a lepton is *slepton*,

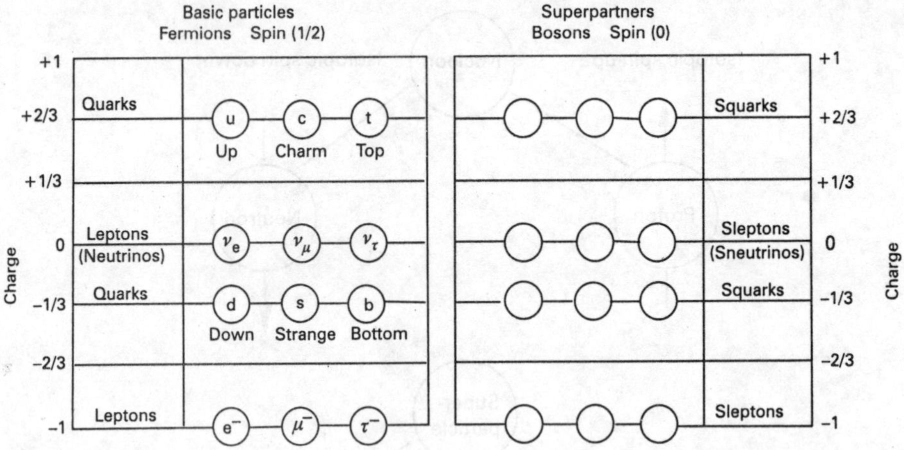

Fig. 4.2 *This figure shows on the left the fermions associated with the SM. On the right are shown the superpartners permitted by SUSY. Note that for reasons of simplicity, the antiparticles of quarks, etc., and their superpartners are not shown.*

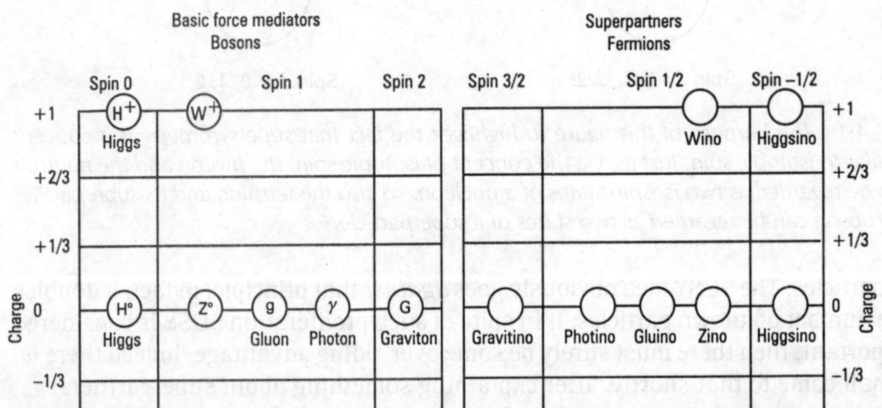

Fig. 4.3 *This figure is similar in spirit to the pervious one and is concerned with bosons of the SM and their superpartners.*

and so on. Figures 4.2 and 4.3 summarise some of the important properties of the superpartners.

Do these superpartners really exist in Nature? Have they been found? The answer to the first question is that physicists now believe that superpartners *do* exist. And the answer to the second question is that they have so far not been found even at the highest energies available. The question may then be asked: "Then why on earth invent these particles in the first place? And why have they not been found, if they actually exist?"

Let me deal with the second question first. To start with, suppose that the slectron had the same mass as the electron. Then a selctron could bind to a proton to form a new kind of atom. For a moment, let me call this *shydrogen*. This atom of shydrogen would have very different properties compared to the hydrogen atom we are familiar with. Can you guess why? Very simple; the electron being a fermion has to respect the Pauli principle, and that in turn results in a very particular type of energy levels. In shydrogen atom the negative charge is carried by the selectron which is a boson. So, it does not have to respect the Pauli principle; as a consequence the energy level structure would be entirely different. Such an atom has, however, so far not been observed. One therefore concludes that the selectron must have a mass very different from that of the electron.

OK, but why have an selectron in the first place? Ah, that is the whole point about invoking SUSY! You see, earlier we discussed physics at the energy scale of electro-weak interactions and there we had to deal with W^+, W^- and Z^0. These have masses around 80–90 MeV. In the same manner, it is reasonable to expect that when we go to Planck energies, we would have to come to terms with new particles. According to the argument just made, the superpartners of the W and the Z, namely the *wino* and *zino,* with their vastly different masses could very well be these exotic particles operating at Planck energies. In other words, when we want to go far beyond GUT, we would have to encounter new particles with masses of the order of the Planck mass; SUSY offers a natural and elegant way of creating space for these additional particles.

But wait a minute! Aren't we missing something here? Yes, your guess is correct. *We cannot have perfect SUSY!* If we did, then the wino would have the same mass as the W. So, if the wino is to have a very different mass, *then SUSY must be a broken symmetry.* This is important.

In a nutshell, SUSY provides a natural symmetry between bosons and fermions in particle physics. At the same time, SUSY is not perfect; it is a broken symmetry. This allows for the mass difference, in fact the masses appear on two vastly different scales of energy, and that is exactly what we want.

This is not all; SUSY removes the exclusiveness of bosons and fermions, that is to say, it allows a fermion to transform into a boson and vice versa. There is something more, and that in fact is the most exciting aspect of SUSY. By the way, when we talk of SUSY we are talking of a local symmetry and not a global symmetry—I have already explained the difference between these two types of symmetry in Volume 1.

Consider now Fig. 4.4. It shows a fermion to which a SUSY transformation is applied. The fermion then becomes a boson, and this is exactly what SUSY is supposed to do. In addition to this change of species, there is an interesting and a very surprising outcome—the *position of the particle changes*. This translation in space is very important because it has deep implications.

SUSY is an internal symmetry, i.e., the transformation is applied in an internal space that has nothing to do with spacetime—I have discussed internal transformations in Volume 1. Nevertheless, the repeated application of SUSY transformation moves a particle from one point to another in spacetime, and this is what Fig. 4.4 attempts to illustrate. In other words, *SUSY transformations generate*

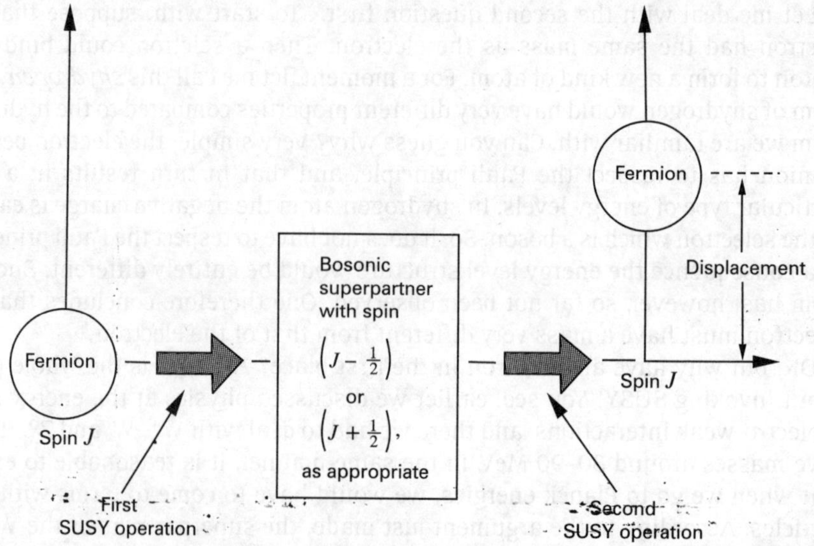

Fig. 4.4 *The purpose of this figure is to draw attention to the fact that repeated application of the SUSY operation (which, remember is in internal space) causes a displacement or translation in spacetime. The big arrow represents a SUSY transformation, and when applied twice one finds that the original particle has actually moved in space. This result is of profound significance since it connects internal and external symmetries.*

transformations in spacetime or transformations associated with the Poincare group. Why is this important? It is important because local Poincare invariance gives rise to gravitational force. This, incidentally, was Einstein's discovery. We now see that SUSY offers an attractive prospect for linking the internal symmetries of particles (that dominate GUT) with Poincare transformations that form the basis of general relativity and gravitation. This is a very important alliance, the linkage between internal space and external spacetime. Later we shall see how all this helps us on to supergravity.

4.3 Quantum gravity

When Maxwell formulated his famous electromagnetic theory towards the end of the nineteenth century, no one even dreamt that there would be such a thing as quantum theory. Even after the birth of quantum mechanics it took about twenty years for a proper union between electromagnetic theory and relativistic quantum mechanics. I am alluding here to the development of quantum electrodynamics which I have discussed in ***QR II***.

Turning next to the weak and the strong forces, it was recognised right from the start that these must necessarily be described in the language of quantum mechanics. We have already seen in Volume 1 how the electric and the weak forces were unified and how later a quantum theory of the strong force, i.e., quantum chromodynamics was also developed. Thus, by the end of the twentieth century, three out of the four fundamental forces of Nature had been given a proper quantum foundation; the lone exception was the gravitational force, the first of the four fundamental forces to be discovered.

There is a basic difference between the electromagnetic force, the weak force and the strong force on the one hand, and gravitational force on the other. The first three *act* in spacetime, that is, spacetime acts as a backdrop for the events governed by these three forces. Gravity is quite different. It is *not* a force *superimposed* on a passive spacetime background. Rather, as we know from Einstein's General Relativity, *gravity is a distortion of spacetime itself.* Einstein's theory was a classical theory. *Quantum Gravity* is the name given to the theory that would unite Einstein's classical theory of gravitation with quantum mechanics.

Till almost the end of the twentieth century, physicists managed to get along without quantum gravity. How come? The answer is simple; quantum effects in gravity become important only when one goes to extremely high energies like 10^{19} GeV or, equivalently, to very small scales of length like the Planck length or 10^{-33} cm. Here we should recall that Maxwell's theory of electromagnetic radiation could get along without quantum mechanics till one started applying it to atoms. Once one started investigating physics on scales of length 10^{-8} cm and below, a marriage between quantum theory and Maxwell's theory of radiation became an absolute necessity. Likewise, quantum gravity becomes a must when one wants to discuss physics on a length scale of the order of Planck length.

You might argue as follows: "There is as yet no proper GUT. First let us get that settled. After that we could seek a SUPER GUT which duly incorporates a proper theory of quantum gravity. Till we are ready with SUPER GUT, where is the hurry?" This is certainly a point of view. However, even without SUPER GUT (if you want to call the SUSY-based GUT that), the question of unifying quantum theory with classical gravity is a sufficiently challenging dare. Thus it is that the subject of quantum gravity has attracted many ambitious investigators, even as unconquered frontiers always beckon brave explorers. Has any breakthrough been achieved? Not yet. Nevertheless, we should get acquainted with the complexity as well as the fascinating aspects of the problem.

Now when we talk of gravitation on scales of the Planck length, we cannot help making a reference to the Universe itself. Why? Because such physics must have applied to the Baby Universe almost immediately after the Big Bang that gave birth to the Universe. Keep this in mind. By the way, don't worry if you haven't heard of the Big Bang and don't quite know what it is all about. We would be spending a lot of time on that topic later.

Let us now go back to the sixties when Roger Penrose (his name must be familiar from many of the earlier volumes) and Stephen Hawking (see BOX. 4.1) noted for the first time that Einstein's (classical) theory of gravitation is inadequate in certain limits. More specifically, Einstein's theory predicts infinite matter/energy density under certain circumstances. In mathematics we take infinity in our stride—no problem there—conceptually or otherwise. In physics as we normally know it, we often encounter infinity in certain limits but do not bother about it; we simply say, "Oh, we will never reach that limit," and forget about it. Consider, for example,

Box 4.1

This is the story of Stephen Hawking, born in Oxford in 1942, exactly three hundred years after Galileo died. Interestingly, Newton was born in December of the same year as Galileo passed away. I am mentioning these names because the work that Hawking has done has a bearing on the discoveries made by these great seventeenth century pioneers. In addition, Hawking's fame is rivalled only by that of Einstein, who too had much to do with gravitation.

In school, Hawking hardly showed any indication of the great things he was to achieve later. Where marks and studies were concerned, he was just an average student. Nevertheless, he was very creative and spent a lot of time building working models of trains, boats, planes, etc. He also invented many elaborate games. Later he said, "I think these games, as well as the trains, boats, and airplanes came from the urge to know how things worked and how to control them. Since I began my Ph. D, this need has been met by my research into Cosmology. If you understand how the Universe operates, you control it in a way."

Hawking wanted to specialise in mathematics in the last couple of years at school, where his maths teacher had inspired him to study that subject. However, Hawking's father was strongly against the idea, and the poor boy was made to study chemistry instead. This refusal of the father had a telling effect, as I shall point out later.

After school, young Stephen went to Oxford in 1959, where his parents had studied earlier. There Hawking studied physics but did not do too well in the exams, blaming it all on the intellectual ambience. He says, "The prevailing attitude in Oxford at that time was very anti-work. You were supposed to be brilliant without any effort, or accept your limitations and get a fourth-class degree. To work hard to get a better class degree was regarded as the mark of a grey man—the worst epithet in Oxford vocabulary."

From Oxford Hawking moved to Cambridge in 1962 and started to work on Cosmology. In December of that year Hawking came home feeling not very well. His mother noticed that he was somewhat clumsy, and a medical checkup followed. The test revealed that young Hawking was suffering from *amytrophic lateral sclerosis*, known as motor neuron disease in England and Lou Gehrig disease in America. This disease attacks nerve cells that control motor activity, but does not affect brain function. The body quickly atrophies, and paralysis of the limbs sets in. This was a shock, especially as the doctor who examined him 'washed his hands off' as Hawking says, and gave him just two years more of life.

Depression set in, naturally, but Hawking did not try to escape via drugs and alcohol as men of lesser courage would have done. Instead, he sought refuge in Wagner's operatic music. While he was in hospital for tests, he saw a young boy die of leukaemia. He says, "It was not a pretty sight. Clearly there were people who were worse off than me. Whenever I feel inclined to be sorry for myself, I always remember that boy."

Hawking pulled himself out of depression, and back in Cambridge he found that he liked research. He says, "I started working for the first time in my life (a reference obviously to the socially-enforced inactivity in Oxford!). To my surprise, I found I liked it."

Contd...

BOX 4.1

...*contd.*

We shall discuss in a later chapter some of Hawking's contributions. To get on with the story of his life, despite his affliction, Hawking got married to a lady who was devoted to him and gave him much support and help. By 1982 his expenses mounted—medical bills and money for educating his children were the main cause. He decided to write a book in order to make some money. Cambridge University Press came to him with a suggestion that he write a popular book on Cosmology. They also offered what by the scientific publishing business was a generous advance. But Hawking was looking for much more money – he needed it – and so he chose another theme. Thus was born the famous book *A Brief History of Time*, to which I have already made a reference in ***A Hot Story.***

By 1984 a draft of this book was ready. At that time Hawking was in Geneva, visiting CERN. He became sick and was rushed to the hospital. This is what he says about that: "The hospital in Geneva suggested to my wife that it was not worth keeping the life-support machine on. But she would have none of that. I was flown back to Cambridge. There, a surgeon called Roger Grey carried out a tracheotomy. That operation saved my life but took away my voice."

Double difficulty now, no muscle movement and no voice. Yet, Stephen Hawking remains undaunted and does world-class work—that is what spirit and courage are all about. Hawking's voice problem has been solved by a California programmer who supplied a computer program using which Hawking makes sentences on a computer that is fitted to his wheelchair. The voice-synthesizer built into the computer then 'speaks out' the sentence. A hard way to lead even a plain and ordinary life, but you know what, Stephen Hawking gives lectures and interviews! Amazing, isn't it?

Getting back to the book I mentioned earlier, it proved to be a runaway best-seller, has been reprinted any number of times, and has made a place for itself in *Guinness Book of Records* as the book that has remained for the largest number of weeks on the best-seller list. Naturally, all kinds of honours have been showered on Hawking.

An interesting aside. Though with a talent for mathematics, because of his handicap Hawking could not do extended derivations of the kind that Chandrasekhar delighted in. He therefore chose problems that did not require much written mathematics, but a lot of conceptualisation, visualisation and ingenuity of course. About all this Hawking says, "I have to be able to visualise a problem. There are plenty of people who can manipulate equations better than me, because it is difficult for me to write them down. Even if I do, I don't have much intuition about equations. I do, however, have intuition about geometry." So there it is, a handicap converted to advantage. People with physical handicaps do this all the time, the blind, the deaf and so on.

Hawking paid a brief visit to India in 2001; he was an instant hit—I am sure you remember that. His work is as profound as that of Einstein, which is most astonishing considering his monumental handicaps. Thanks to his extraordinary scientific contributions, it is no wonder that he has been appointed to the Lucasian Chair in Cambridge University. This post created for Newton remained vacant after him for three hundred years till Hawking came along! Truly, Hawking's life is a profile in courage and determination.

Coulomb's law describing the electrostatic attraction between two unlike charges (see Fig. 4.5). When the distance r becomes zero, the attractive force becomes infinity. We do not bother about this. We simply say, "Well, this law is no good anyway when we go to such small distances. We need quantum electrodynamics." In quantum electrodynamics there were infinities to start with, but these could not be brushed aside like we did in the case of simple electrostatics. As you will recall (from ***QR II***), it took a lot of effort and genius to tackle these (unwanted) infinities.

The infinity arising in Einstein's theory of gravitation is even worse and cannot be dismissed all that easily. As someone said, if the infinity in electrodynamics is like dealing with (1 + 1 + 1 + 1 + ...), the infinity in gravitation is like (1 + 2 + 3 + 4 + ...). That ought to give you some idea of how much more difficult the problem is. When, for example, we consider the Big Bang—and we shall be doing that soon—we have to ask ourselves, "What was the Universe like at the instant of birth?" The answer that one usually obtains is: "It had zero size and infinite density." "You mean we have to deal with infinity?" "Yes, infinity!" "But we can't have that in Physics!" Well, that is one of the reasons for seeking quantum gravity. With quantum effects one could possibly blur the Big Bang infinity and perhaps fix the problem—that is the hope and the topic will find a mention later.

OK, so what has happened? Has the Holy Grail been found? Not yet – only lots of problems! I can't go into that story for the simple reason I do not even know

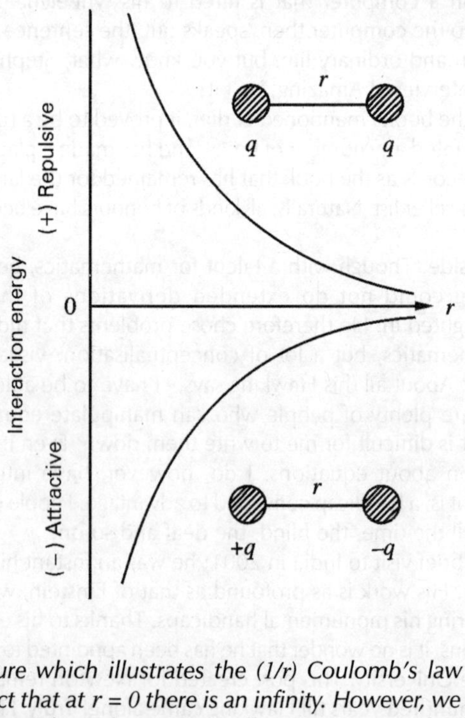

Fig. 4.5 *This figure which illustrates the (1/r) Coulomb's law is included just to draw attention to the fact that at r = 0 there is an infinity. However, we do not bother about it for an important reason. When r becomes small, we cannot use classical physics anyway; so the singularity at r = 0 becomes irrelevant.*

how to describe these complex problems! However, let me mention just one of these problems. It turns out that when we are considering a situation where spacetime itself is small—and that is what we have to do when we are discussing the state of the Universe immediately after the Big Bang—*spacetime itself becomes grainy and starts fluctuating!* This is a bizarre situation such as never encountered before in physics. No one knows as yet how exactly to handle this most unusual complication. But of one thing people are sure: If Einstein's theory and quantum theory are to be unified, *then both must give way,* that is, one may have to modify the very foundations of quantum mechanics as also the foundations of the classical theory of gravitation. Thus far, in all the explorations in physics, even right down to GUT, the foundations of quantum mechanics (as opposed to its philosophical interpretation) have remained unchanged from the nineteen thirties to the end of the twentieth century. So you see, a tremendous breakthrough is needed in this case. Perhaps such a breakthrough might come from an unexpected quarter, and indeed there is sufficient reason to believe that this might happen; more about that later.

Let me wrap up this brief discussion on quantum gravity by saying that everyone realises that without a proper marriage between quantum mechanics and the classical theory of gravitation there cannot be any question of discussing physics at the Planck scale of length. Such a unification between quantum mechanics and gravity has yet to take place since very great difficulties have yet to be overcome. But there is hope that the problems will get solved, though not necessarily in the manner we might presently think.

4.4 Supergravity

Supergravity is the name given to a theory of gravitation that rests on supersymmetry. It is thus quite distinct from quantum gravity that I discussed in the preceding section, where a seemingly 'straightforward' unification of quantum mechanics and general relativity is attempted at without bringing in any particle physics. By contrast, supergravity seeks to extend SUSY to take account of gravitation also, by the explicit use of the fact that in SUSY internal symmetry gets related to Poincare symmetry in a natural fashion as just explained. Figure 4.6 illustrates the structural relation underlying the two types of theories.

Has supergravity type of theory achieved any breakthrough that was not possible with quantum gravity? The short answer to that is: Not yet. However, various models of supergravity theories that have been explored have thrown up many interesting facts. The first one to be investigated considered just gravitons (associated with gravitation) and gravitinos (associated with supersymmetry). This is clearly unrealistic since all the other basic building blocks are left out. Nevertheless, this simplified model was enough to surface the problem of infinities that plagues all field theories. And then, after much investigation, it became clear that in a certain class of theories of supergravity called *Extended Supergravity Theories*, some cancellation of infinities is possible.

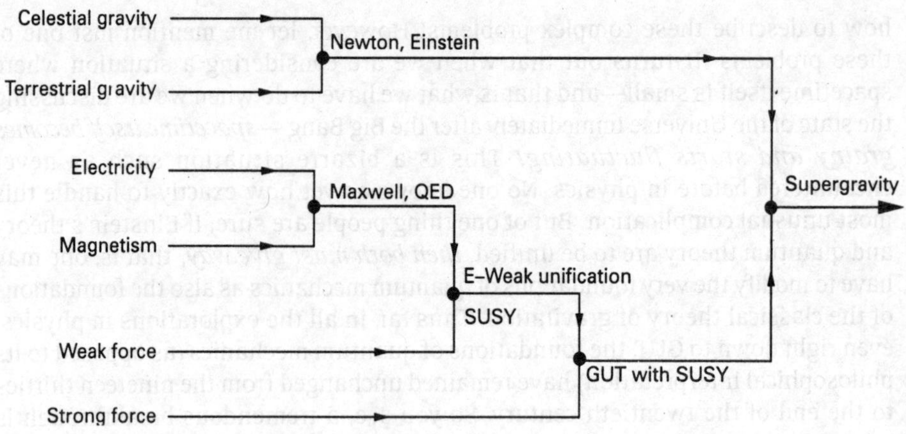

Fig. 4.6 *This highlights how supergravity emerges when all the preceding steps of unification have been completed.*

Partial success in dealing with infinities is no doubt a plus but many problems remain. For example, the particle masses all come out as zero! We can't have a theory predicting zero mass for the electron, can we?

At this point you might wonder: "Why on earth do physicists waste their time chasing non-working theories?" Well, these daring souls are not exactly wasting their time. Almost invariably, breakthroughs in science come only when a lot of legwork has been done. Often, this legwork is very tiring and demanding. Sometimes it leads to disappointing ends. Many snobs ridicule such work as low-grade, which is unfair. Could Kepler have made his monumental contribution but for the painstaking astronomical documentation produced by Tycho Brahe? The history of physics, and indeed of science, is full of stories of people who patiently collected data and explored a variety of ideas that facilitated a breakthrough later. These unsung and unhonoured heroes must also be saluted. In the same way, the so-called failure of these people knocking away at quantum gravity or supersymmetry, or for that matter supergravity, has in its own way contributed to progress. Remember, failure is often the stepping-stone to success! We have seen earlier in Volume 1 how a large number of people had to struggle and fail before Salam and Weinberg got the clue as to how to unify the electromagnetic and the weak forces. The moral of it all is: they also serve that valiantly try but do not quite make it.

4.5 Summary

- This chapter offers the first glimpse of the territory that lies beyond the early GUT type of theories.
- Extensions have been made in three directions, of which the first is SUSY.

- Attempts have also been made to bring QM and GR together under the umbrella of quantum gravity.
- Finally, SUSY has been extended to incorporate gravity via supergravity.
- Based on these pioneering foundations, much more has been done. That story starts in the next chapter.

5

IS GOD PULLING STRINGS?

This chapter discusses what is called the String Theory, wherein elementary particles are not viewed as point particles (as has been implicitly done all along) but as tiny strings, perhaps of length of the order of Planck length and capable of vibrating in many modes. This approach has turned out to be a whole new ball game, as they say.

In physics, there is a long tradition of using the particle concept while dealing with matter. Newtonian mechanics, for example, is based almost entirely on the concept of particles. Indeed (for reasons that you ought to know!), we can get away with the particle concept even while dealing with many problems in astronomy.

Later came the kinetic theory of gases (see ***A Hot Story***), and once again the particle concept was used. No one bothered about the fact that the atom could have a finite size; instead, one just treated the atom as if it was a geometric point and regarded the trajectories it traced as ideal geometric lines. The remarkable part of it all is that one could get away with such sweeping and simplifying assumptions!

When it came to atomic physics, however, one could no longer treat the atom as a point, but where the electron was concerned it was implicitly regarded as if it were a point. Of course, in quantum mechanics we have the duality concept, fuzziness and all that, but at the back of the mind the point particle concept remained as it were. I mean, when one talks of the position of an electron, one implicitly takes it to be a particle of zero size.

I shall not get into a debate on this matter; I know you can pick many holes in my remarks! All I want to do is to draw your attention to the fact that since approximately the past twenty years, many particle physicists have started viewing elementary particles as *strings rather than point particles*, which is what they were implicitly doing earlier. The string idea is thus a *major* departure from the traditional point of view. As I shall shortly explain, these physicists *explicitly* invoke the idea of strings. In fact, such people belong to a club of their own—they are string theorists and not just particle physicists!

Like many concepts in physics, the string idea came in almost unnoticed. Somewhere in the seventies a physicist named Veneziano, then at CERN, simply *guessed* a formula that was able to explain many features of hadron resonances. Subsequently, Nambu of Chicago (remember him? He makes his appearance in Volume 1) showed that using Veneziano's formula is equivalent to treating hadrons as different modes of vibration of a string. You know that a string can vibrate and has many harmonics—this is something that is taught to us in elementary physics. Nambu argued that if Veneziano's idea is taken seriously, then it would mean that the different hadrons were just different harmonics of the string. In Volume 1 we had already encountered an analogous idea with respect to particle resonances—I mentioned then that many of the elementary particles were just excited states of the nucleon, the pion, and so on. Nambu's idea is something similar.

Sometime after Nambu gave his clarification, Ramond and Schwarz extended the original work to include fermions also—the original model of hadron resonances was applicable only to bosons. This new theory was called the *Spinning-String Theory* and was in a manner the precursor of SUSY. This new theory did not hit the headlines and lay ignored for many years until it was revived in a new form by Schwarz and Green in 1980. That was when the String Theory as we now know may be said to have been born.

Strings are the elementary objects in string physics and can be of two kinds—open or closed. We know from special relativity (see ***Relativity***) that a moving particle traces a world line in spacetime (see Fig. 5.1 *a*). In a similar manner, when the string sweeps through spacetime it produces a *world sheet* (see Fig. 5.1 *b* and *c*).

Next we consider the action principle as it operates in string physics. You will recall that in classical physics we have the Principle of Least Action, a powerful principle that attracted Feynman so much. What this principle tells us in effect is that when a particle goes from one state to another, it chooses a trajectory for which action is a minimum. In practical terms, the trajectory is a

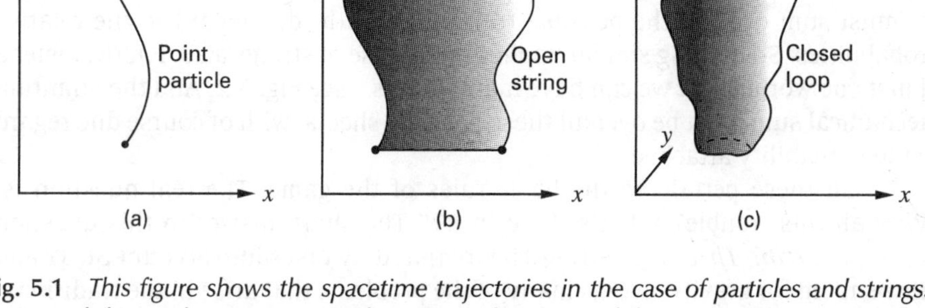

Fig. 5.1 *This figure shows the spacetime trajectories in the case of particles and strings. In (a) and (b), we have one space dimension plus time, while in (c) we have two space dimensions plus time. What is a line in (a) becomes a sheet in (b) and a tube in (c). This has important implications as we shall see soon.*

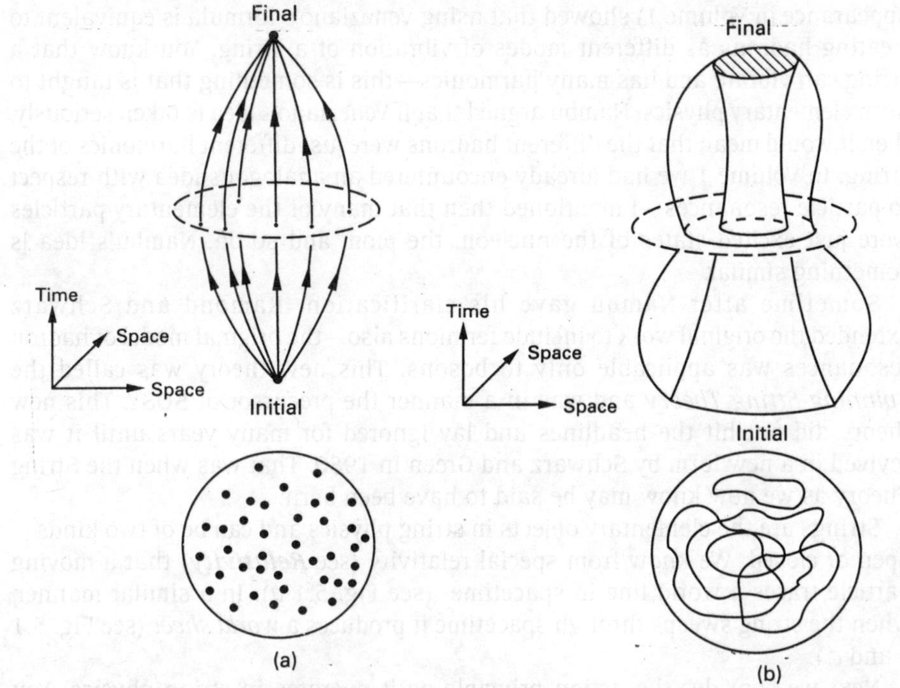

Fig. 5.2 *This figure seeks to illustrate the implications of integrating over all possible trajectories between the initial and the final states. Let us say we have two space and one time dimension. We now take the cross-section across space at a particular time. If we are dealing with world lines, we would see, as in (a), a pattern of scattered points in the cross-section taken. If we are dealing with world tubes, we would see a superposition of closed curves of arbitrary shapes as in (b).*

line of minimum length in spacetime, joining the initial and the final states. Of course, when one goes over to quantum mechanics one has to give up the idea of a privileged trajectory and allow for all possible trajectories, with, however, different probabilities attached to them. And QM also tells us that when evaluating the final outcome or the transition probability from the initial to the final state, we must sum over all the possible trajectories with due regard to the relative probabilities. Something similar applies in the case of strings also. Firstly, instead of just one world sheet we can have many sheets – see Fig. 5.2. And the quantum mechanical sum must be over all these possible sheets, with of course due regard to the probability attached.

OK, all these pertain to the basic rules of the game. The real question is: "Why all this trouble? What's there in it?" The short answer to this question is: "*Super String Theory*, or String Theory that duly takes into account SUSY and allows room for gravity, avoids many of the problems that arise in the traditional marriage of quantum mechanics and classical theory of gravitation."

Another great plus—and in a measure this is related to the point made above—is that infinities are less of a problem in string theories than in point-particle

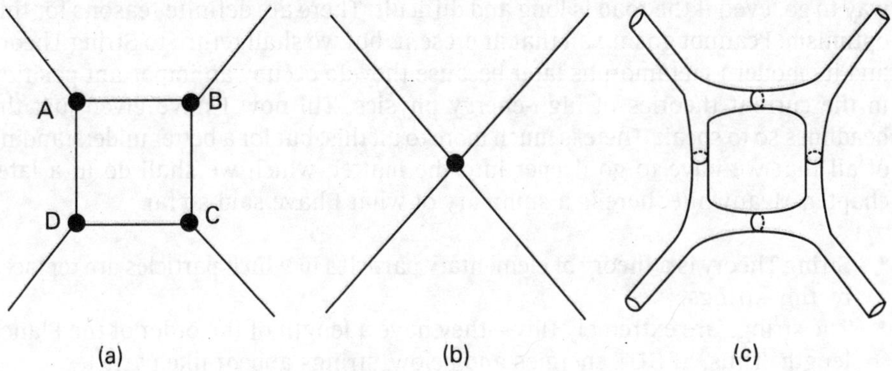

Fig. 5.3 *This figure seeks to explain how the problem of singularities is avoided when one goes from particles to strings. (a) shows a Feynman diagram with four vertices. When integrating, one must allow for all possible locations of the four vertices; in particular, one would have to consider a situation as in (b) where the vertices overlap. It is this situation that leads to singularities. When one deals with strings instead of particles, the Feynman diagram takes on the appearance as in (c). As is readily seen, the finite extension of the string smears out the singularity.*

theories. This is really a very technical matter, but the hand-waving way in which this complicated fact is usually explained is given in Fig. 5.3. Recall first that in field theory when one considers the transition from an initial state to a final state, one must consider all possible physical processes that could describe such a transition. In the language of Feynman diagrams, one has to deal with sums, as I have explained in detail in ***QR II***. As you will recall, sums such as these can lead to infinities, and this was the plague of field theories. Consider the diagram in Fig. 5.3 (a). While evaluating the sum we must allow for all possible locations of the vertices A, B, C, D. Trouble comes from a situation as in Fig. 5.3 (b) where the spacetime events A, B, C, and D of (a) are no longer distinct but coincide. This in turn leads to a singularity in the integrals to be evaluated. But when points become strings this troublesome situation is avoided, as can be appreciated from Fig. 5.3 (c). This is one of the bonuses that comes from string theory. Once singularities do not occur when making sums over Feynman diagrams, one no longer has any problems with infinities.

Two questions: First, if elementary particles are indeed strings and not points, how come we have got away so far with the notion of a point particle? The answer to this is that the string length is so small that strings appear as strings only when one goes to Planck kind of energies. At the kind of energies studied thus far in high-energy physics, strings *do* appear to behave as particles. I mean, it is like saying that an atomic nucleus appears as a point until we start doing physics in the low MeV energy regime. That in brief is String Theory's answer to the first question.

Second question: Has String Theory made any predictions that can be tested? None at all. Then why are people going gaga over it? Because it definitely holds much promise. Most serious physicists are convinced that this is the way to go, even if the road is long and difficult. There are definite reasons for this optimism. I cannot go into all that at present, but we shall return to String Theory and its modern metamorphs later because they do occupy an important position in the current theories of high-energy physics. Till now I have given just the headlines so to speak. There is much more to all this, but for a better understanding of all that we have to go deeper into the matter, which we shall do in a later chapter. Meanwhile, here is a summary of what I have said so far.

- String Theory is a theory of elementary particles in which particles are replaced by tiny strings.
- The strings are extremely tiny—they have a length of the order of the Planck length. Thus, at GUT energies and below, strings appear like particles.
- Just as particles trace world lines in spacetime, strings trace world sheets in spacetime.
- The strings have a tension which is described via a constant α' that is the analogue of $\hbar$ in quantum mechanics.
- Super string theories have the remarkable property of *predicting* gravity. That is, they automatically require the existence of massless spin-2 particles or the graviton whose field behaviour at long distances is consistent with general relativity.
- The problem of infinities is substantially tamed.
- Super String Theory not only leads automatically to gravity but also provides room for SUSY in a natural way.

If it is all that good, then surely String Theory must be the Theory of Everything! Indeed, that is how it was promoted for a while!! But it did not turn out that way. However, there is a lot of hope that the Super String Theory, especially the exotic extensions of it that have appeared recently, may eventually deliver the goods. More about that later.

6

COSMOLOGY — A FIRST GLIMPSE

Moving from particle physics, we get in this chapter our first glimpse of modern Cosmology, which starts from the time Einstein proposed his theory of gravitation. Einstein applied his theory to the Universe itself, following which it was discovered that the Universe had a definite birth and expanded subsequently. Hubble discovered experimentally such an expansion. Gamow later visualised the state of the birth of the Universe as a tiny but incredibly hot fireball, the Big Bang, in which the very first elements were formed. A natural corollary to this was that accompanying the Big Bang there must have been a big flash of radiation whose remnants must be around even today. That is the cosmic background whose discovery is described here. All this led to many fascinating questions which eventually brought particle physics and Cosmology together, as described in later chapters.

6.1 Structure of the Universe

At last we move from the small to the big; in this chapter we briefly consider some aspects of the physics of the Universe. Now, why are we doing this? The answer is the following: At the present time the Universe is roughly about fifteen billion years old. Radiation from the farthest corners of the Universe as it is at the present time, would take about fifteen billion years to reach the earth if that radiation starts now and there is no further expansion of the Universe—that is how vast our Universe is! Yet, when it was born—we think it was born in a Big Bang, and more about that shortly—the Universe was very tiny. In fact, 10^{-43} seconds after birth (which is the Planck time), the Universe was incredibly small. And the physics that must have operated then must have been string physics or something like that. That is what makes the Baby Universe so exciting to physicists who are interested in the behaviour of matter at scales of the order of Planck length. By the way, now is when you should brush up on the Planck units discussed in Chapter 1. From now on, these units would crop up quite frequently.

Now you might object and say, "Hey, wait a minute! Who knows what happened fifteen billion years ago? How can I be asked to accept a theory and be told, 'You know, this was the physics that worked fifteen billion years ago!'" Yes, how are we to know what happened that long ago? This is where the whole thing becomes like a detective story!

Let us say Sherlock Holmes has been summoned to the scene of a crime. He carefully surveys the scene, perhaps making tantalising comments that leave his friend Watson totally mystified. From the evidence available, Holmes then reconstructs the crime. That is the strategy of elementary particle physicists today. They say: "Listen, if our speculation about physics at small distances is correct, then these are the things that must have happened in the Baby Universe. If the Universe was that kind of a baby then, then it ought to be an adult of this kind now. Will the astronomers and astrophysicists please peer into the sky and find out if the Adult Universe is like what we are saying it would be?" The arguments of the particle physicists are so compelling that many astronomers and astrophysicists are indeed looking carefully at today's Universe and checking out if what they are seeing is consistent with the predictions of particle physics. This is roughly the reason why astrophysicists and particle physicists have recently started hobnobbing with each other.

Keeping that background in mind, let us now have a quick glance at the Universe as a whole and what it has to tell us. Let us start with the objects to be found in the Universe on different scales of length. For dealing with large distances we normally use the light year as the unit of length. Astronomers prefer to use the *parsec* as the unit of distance. One parsec, abbreviated pc, is that distance at which half the diameter of the Earth's orbit around the Sun subtends an angle of one arc second. More explicitly,

$$1 \text{ parsec (pc)} = 3.0856 \times 10^{18} \text{ cm or} \sim 3.26 \text{ light years.}$$

Turning to the contents of and the structure of the Universe, to start with we have planets like the Earth, the Moon, Mars and so on which form a part of the solar system. The Sun which is the focus of the solar system, is a star. Like our Sun there are about at least a billion stars, yes, at least a billion that form a part of our galaxy known as the Milky Way. I am sure you already know all this.

Looking beyond the galaxy, we have what are called the *local groups* (LG). A local group may consist of about 10–50 galaxies. The Milky Way belongs to a group of about twenty galaxies. The Andromeda galaxy, about which you must have surely heard, is a member of the LG to which the Milky Way belongs.

After the LG comes the *galaxy cluster*, a typical cluster having something like a thousand galaxies and a linear dimension of about 5 megaparsecs or Mpc. Next comes the supercluster, which is about ten time larger than a cluster, i.e., it has a linear extent of about 50 Mpc.

Structures of even larger scales are seen in the sky, and recently astronomers have noted that something like filaments of bright objects exist. One of them has even been named the Great Wall! It has an extent of about 60 Mpc × 160 Mpc, and is made up of several superclusters. In short, the sequence of large structures goes something like this (see also Fig. 6.1):

Stars → Star Cluster → Galaxy → Local Galaxy Group → Galaxy Cluster → Galaxy Supercluster → Filaments and Wall → What else? → Edge of the Universe

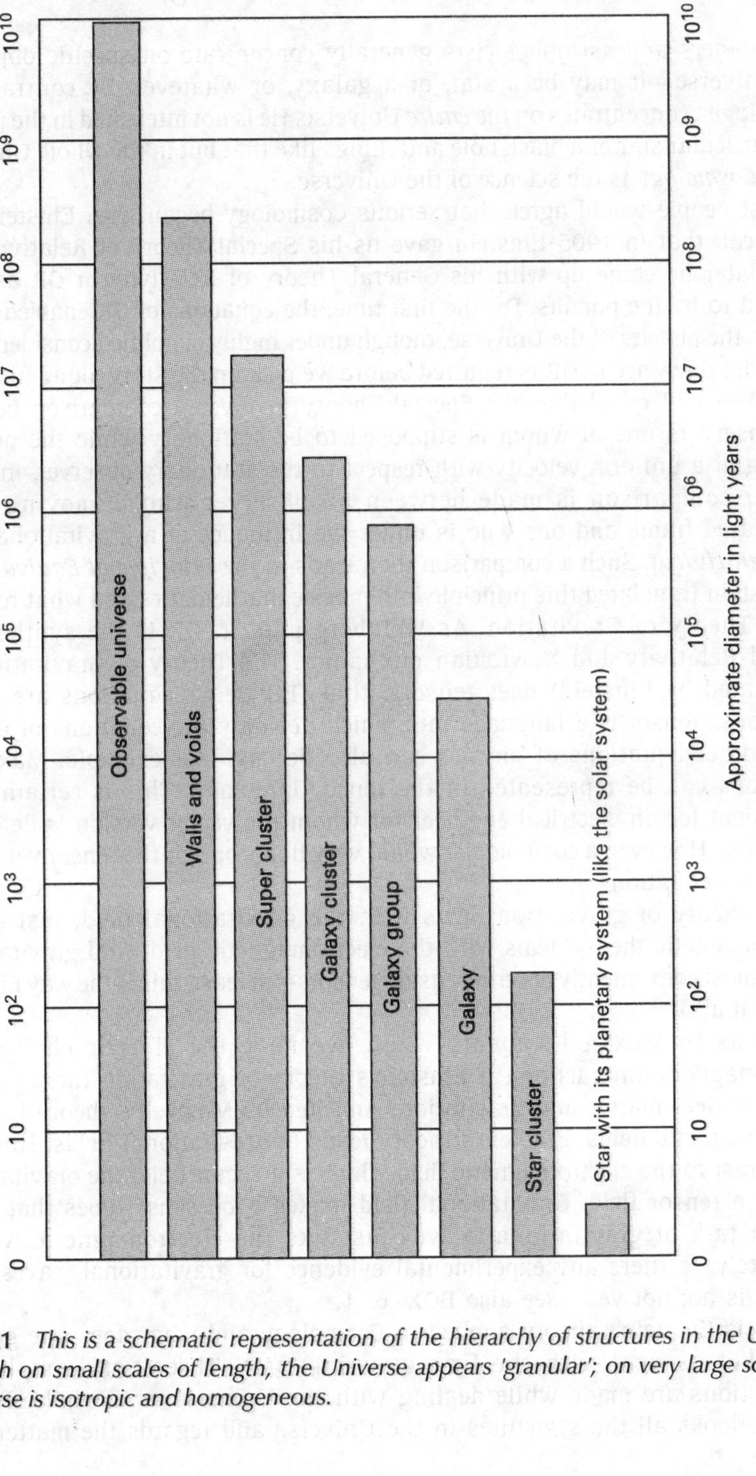

Fig. 6.1 *This is a schematic representation of the hierarchy of structures in the Universe. Though on small scales of length, the Universe appears 'granular'; on very large scales the Universe is isotropic and homogeneous.*

6.2 The Beginnings of Modern Cosmology

Astronomers and astrophysicists generally concentrate on specific objects in the Universe—it may be a star, or a galaxy, or whatever. By contrast, the cosmologist concentrates on the *entire* Universe. He is not interested in the physics of a particular star or a black hole and things like that but in the whole Universe. Thus, *Cosmology* is the science of the Universe.

Most people would agree that serious Cosmology began with Einstein. You will recall that in 1905 Einstein gave us his Special Theory of Relativity. Ten years later he came up with his General Theory of Relativity or GR as often referred to by the pundits. For the first time, the equations of GR enabled one to discuss the physics of the Universe, though under highly simplified considerations.

A brief reference to GR is required before we pick up the story along historical lines. You will recall that the Special Theory involves a comparison between two observers one of whom is supposed to be stationary while the other is moving at a uniform velocity with respect to the stationary observer. In GR, a similar comparison is made between an observer who is moving in an accelerated frame and one who is under the influence of a gravitational field [see ***Relativity***]. Such a comparison then leads to the *Principle of Equivalence.*

Einstein translated this principle into precise mathematics and what resulted is his Theory of Gravitation. As Weinberg puts it, GR is the synthesis of Special Relativity and Newtonian mechanics. The Theory of Gravitation (as formulated by Einstein) uses tensors; thus, Einstein's equations are tensor equations. Tensor is a language into which not only the equations of GR but many other equations of physics can also be cast. For example, Maxwell's equations can be represented in the tensor language. This is certainly not convenient for an electrical engineer for whom the vector version is definitely preferable. However, a cosmologist would very likely opt for the tensor version of Maxwell's equations.

The theory of gravitation deals with the gravitational field, just as the electromagnetic theory deals with the electromagnetic field. And gravitational field is most conveniently described using tensors—at least, this is the way Einstein started it all.

Just as in Maxwell's theory where we have the electric charge and electromagnetic interaction, in Einstein's theory of gravitation there are two basic entities: matter and gravitation. And just as Maxwell's theory leads to electromagnetic fields, Einstein's theory leads to gravitational fields. However, in contrast to the electromagnetic field which is a vector field, the gravitational field is a tensor field. Gravitational field makes good sense; does that mean we can talk of gravitational waves just like the electromagnetic waves? Absolutely. Is there any experimental evidence for gravitational waves? The answer is no, not yet – see also BOX. 6. 1.

How is Einstein's theory applied to Cosmology and what does one get out of it? Let us deal with the first question first. Three very simplifying assumptions are made while dealing with the Universe as a whole. Firstly, one overlooks all the structures in the Universe and regards the matter in it

BOX 6.1

When Newton discovered the law of gravitation, the question arose: Suppose a mass m_1 separated from another mass m_2 by a distance say r is moved, and the distance becomes $r + dr$. How would mass m_2 know about the movement of m_1, and how long would it be before the news from m_1 reaches the other particle m_2? It was argued in those days that the transmission of information about the movement would be instantaneous, and this phenomenon was referred to as action at a distance. But when it became clear from the Special Theory of Relativity that information could not travel at a speed greater than that of light, action at a distance automatically got ruled out. Where electromagnetic interactions were concerned, Maxwell's theory had already made room for a field, with waves of that field acting as messengers. Was there such a field where gravitation was concerned? That question was answered decisively by Einstein's Theory of Gravitation; yes there was such a field, there were waves of this field and these travelled with the speed of light.

OK, in that case, how about detecting those waves? Not easy since gravitational waves interact rather feebly with matter. Nevertheless, in the late sixties of the twentieth century, J. Weber, set about trying to detect these waves. He even reported detecting these waves but since subsequent experiments could not verify his results, his claims were not accepted. In a later chapter I shall describe attempts being made currently to detect gravitational waves. But in the meanwhile, Nature itself has provided some very beautiful evidence, though indirect perhaps.

In 1979, a pulsar was detected that was orbiting around another compact star, an invisible one, with a period of about eight hours. The orbital period of the pulsar was found to be steadily decreasing. The decrease is very slight, the accumulated decrease being only about 0.000414 seconds in four years! The most likely explanation for the observed decrease is that the binary system is losing energy through the emission of gravitational waves, causing the two stars to slowly spiral towards each other. This spiralling effect would in turn produce a decrease in the orbital period of the pulsar. Thanks to this discovery, there is no doubt anymore about the existence of gravitational waves. All that remains is to detect them actually here on Earth.

as a homogenous fluid. This is, if you like, a cosmological fluid! Next, one supposes that world lines of particles in the Universe do not cross each other; rather, they follow non-intersecting trajectories as illustrated in Fig. 6.2. This assumption is due to Weyl, and for cosmologists particles often means galaxies! Thirdly, one makes a hypothesis that is called the *Cosmological Principle.* Briefly, this principle states that the Universe is homogeneous and isotropic. What it means is that given any position in the Universe and any specified direction of observation, the large-scale structure of the Universe appears the same.

Einstein was, understandably, the first one to use his field equations to examine what they had to say about the Universe, together with the assumptions just mentioned, in particular the Cosmological Principle. I think Einstein carried out this investigation somewhere around 1917 or so. I shall discuss aspects of Einstein's work shortly, but here let me mention something else that is important. To his

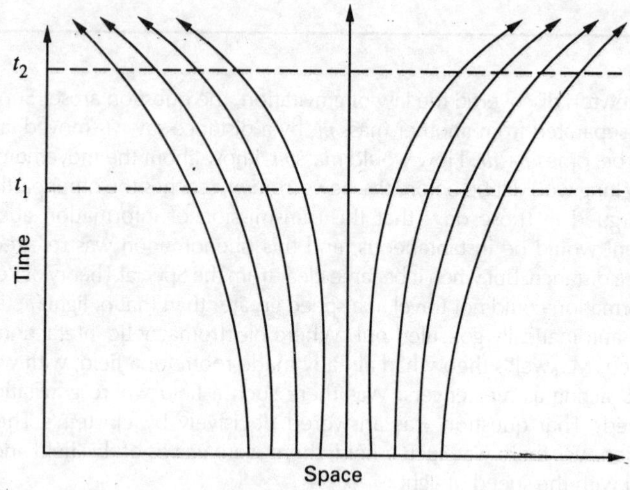

Fig. 6.2 *This figure illustrates Weyl's idea of non-intersecting world lines, each line being due to say, a galaxy (remember, in Cosmology, a galaxy is often treated as a particle!). A bundle of lines as above allows one to identify different epochs of time and also discuss changes as one moves from one epoch to another.*

great astonishment Einstein found that his equations predicted a Universe that was changing in size. Einstein was simply stunned by the discovery. At that time no one ever dreamt that the Universe was changing in size. Somehow, everyone tacitly assumed that the Universe always existed, and that it was ever in a steady state with neither growth nor shrinkage. Now why on earth did everybody believe this? I cannot say; perhaps it was the feeling in the air, influenced possibly by religious proclamations.

Anyway, to get on with the story, when Einstein discovered that his equations predicted a Universe of changing size, he immediately assumed that there was something wrong with his equations. He then doctored them so that they predicted a steady Universe. This he did by introducing a constant known as the *cosmological constant,* denoted by the symbol Λ. This constant applied the brakes and held the Universe back so to speak, preventing it from changing its size! Einstein felt happy, and there matters rested for many years. No one paid serious attention to either Einstein's discovery or his attempt to 'fix' the problem. Dirac once remarked that his equation was smarter than himself. Einstein discovered that the same applied to his field equations also!

6.3 Enter Friedmann and Le Maître

From 1922 to 1924, a Soviet mathematical physicist named Alexander Freidmann from Leningrad [see BOX 6.2] carried out an independent analysis of Einstein's equations of Cosmology. A useful way of understanding Freidmann's work is to start with the concept introduced by Robertson and Walker in the mid-thirties of the twentieth century. They noted that spacetime can be spherical, flat,

BOX 6.2

Alexander Friedmann was born in 1888 in St. Petersburg in Russia. In school, Freidmann was brilliant in physics and mathematics but he also became obsessed with politics! (Those were the days when the tyranny of the Russian Czars was severe. Later, this political fervour would make Friedmann become associated with the communist movement.) In 1906 Friedmann entered the St. Petersburg University. Initially he worked in mathematics. In 1913 there came a change forced by the job he took up. He got an appointment in the Geophysical Institute in St. Petersburg; this made him shift his interest from mathematics to theoretical meteorology, and also spurred his interest in hydrodynamics. In fact, during this period he often flew in balloons to learn about the atmosphere, and in the process achieved reputation as an airman. Incidentally, throughout his life, Friedmannn retained a strong interest in hydrodynamics and meteorology.

It was at this stage of Friedmann's life that the First World War (1914–1918) broke out. For a part of this period Friedmann was an instructor for airmen in the Military School in Kiev. In 1917 he was transferred to Moscow. Shortly after this he found himself in Perm University where he became a Professor in 1920.

Friedmann now became interested in relativity and Cosmology. As one Russian scientist remarked then, "Friedmann is studying Weyl, and we shall soon know what relativity is!" In the main chapter we mention something that is referred to as the Robertson–Walker metric. Actually, this was independently discovered earlier by Friedmann, but as it often happens, the names of Western discoverers get etched in history even though their discovery comes later.

With the metric on hand, Friedmann started investigating the fate of the Universe. In 1922 Friedmann sent his paper to the German journal *Zeitschrift fur Physik*, perhaps the most popular physics journal of the twenties of the 20th century. It appears that the Editor sent the paper to Einstein for his views, not surprising since the paper dealt with matters on which Einstein himself had published earlier. Einstein did not like Friedmann's findings according to which the size of the Universe changed with time. The paper was put on hold because the Master did not approve! The well-known Russian physicist Fock writes that Einstein, "with somewhat superior air, said that the results of Friedmann seemed suspect to him, and that he had found an error the correction of which reduced Friedmann's solution to a steady-state Universe." This of course is what Einstein himself had done earlier; he had obtained a solution that showed the Universe to be changing with time.

News of Einsteins' adverse comments reached Friedmann, and on December 6, 1922, he sent a letter to Einstein in which he wrote: "Much-esteemed Professor! From a letter of one of my friends who is now abroad, I had the honour to find that you sent to press in Vol. 11 of *Zeitschrift fur Physik* a short note where you point out that if all the assumptions made by me in my paper *On the Curvature of Space* are adopted, then from the world equations derived by you it would immediately follow that the curvature of the Universe is a quantity independent of time. ...Bearing in mind the definite interest in the possibility of existence of a non-steady state Universe, I venture to submit to you my calculations for checking and review. .... I earnestly ask you to answer my letter, although I know you are very busy. If you find the calculations presented

Contd...

BOX 6.2

Contd...

in my letter to be correct, I would ask you not to refuse my request to inform the editorial office of *Zeitschrift fur Physik* about it. It may be that in this case, you would publish a correction to your opinion or provide an opportunity for reprinting parts of my letter."

Months passed. Einstein was busy and travelling. Actually, even by 11th October, Einstein had left on a long overseas tour that took him to Japan. Meanwhile it was announced that he had been awarded the Nobel Prize and he had to miss the function which was held on December 11 in Stockholm! He returned to Berlin only in March 1923. In early May, the Russian scientist Krutkov managed to meet Einstein personally and argue on behalf of Friedmann. This worked, and on 21st May 1923 Einstein wrote to the German journal: "In my previous note, I criticised the above mentioned work (of Friedmann). My objection was based on a calculation error, as I convinced myself with encouragement from Mr. Krutkov and guided by a letter from Mr. Friedmann. I regard the results of Mr. Friedmann as accurate and shedding new light. ..." As Krutkov wrote to his sister, "I vanquished Einstein in an argument about Friedmann. The honour of Petrograd is saved!"

Unfortunately, Friedmann did not live long. In July 1925, he went up in a special balloon up to an altitude of 7400 metres in order to make meteorological observations. Going to that height without oxygen can be hazardous. So it proved. It aggravated his heart condition, already not too good. In September he died, at the age of thirty seven. He had a student named George Gamow who later made Friedmann's purely mathematical model come alive by adding intuitive physics in typical Gamow fashion. But that is an entirely different story.

or hyperbolic. Examples of two-dimensional surfaces of these three types are shown in Fig. 6.3. Robertson and Walker [RW] gave a basic equation for discussing such types of (four-dimensional) spacetime. Actually, this so-called RW metric was discovered by Freidmann himself, but this fact is not known widely outside. Russian scientists are therefore quite annoyed to see the name of Robertson and Walker attached to something that was discovered much earlier by one of their own; but this sort of thing happens sometimes in science. Getting back to the RW equation, there occurs a quantity $R(t)$, known as the *scale factor.* In case spacetime expands, the scale factor would describe that expansion—the meaning of expansion of spacetime is illustrated in Fig. 6.3.

Now the Robertson–Walker metric equation, as it is called, is not *per se* connected with Cosmology. However, if it is combined with Einstein's field equations, together with the three basic assumptions that I mentioned earlier, then one gets an equation for $R(t)$ that essentially tells us what happens to the Universe. It turns out that in the equation for $R(t)$, there occurs a dimensionless parameter denoted Ω that is related to the density of matter in the Universe. Ω can be < 1, $= 1$, or > 1. Correspondingly, there are three possibilities for $R(t)$ and they describe three possible scenarios for the Universe. These are essentially what Friedmann discovered and are shown in Fig. 6. 4. Thus, one says that depending on the amount of matter

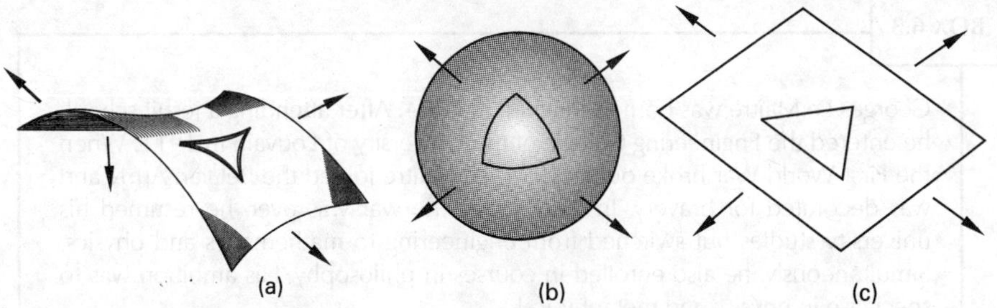

Fig. 6.3 *The equations of Cosmology tell us that spacetime or the fabric of the Universe can be of three types as illustrated schematically above. Shown here are two-dimensional surfaces that are (a) hyperbolic, (b) spherical, and (c) flat. Drawn on these are triangles. As spacetime expands, these triangles get stretched and become bigger and bigger; the arrows indicate expansion of spacetime. By the way, while for a triangle in flat space the sum of the angles adds up to 180 degrees, the same is not true for triangles drawn on spherical and hyperbolic surfaces!*

contained, the Universe can go on expanding endlessly for ever, expand but at increasingly slower rates, or can even start contracting after some stage. This is a very important issue concerning the ultimate fate of the Universe, and we shall come back to it later.

After Friedmann came the Belgian priest-cum-astronomer Le Maître [see BOX 6.3]. In 1927, he presented a theory of an expanding Universe that had a beginning in time. As Le Maître put it, the day of birth was a 'day without a yesterday!' The essence of Le Maître's model is captured in Fig. 6.5 and is more or less a rediscovery of one of the cases discussed earlier by Friedmann. Le Maître was apparently unaware of Freidmann's work. He published his theory in a little-read Belgian journal, and his work lay in oblivion till it was experimentally discovered that the Universe was *actually* expanding.

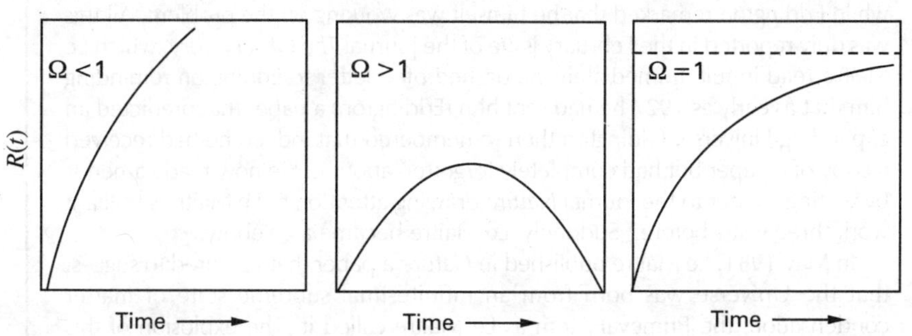

Fig. 6.4 *This figure shows the time evolution of the Universe for different cases corresponding to different values of Ω. When Ω < 1, the Universe is hyperbolic and expands for ever. When Ω = 1, the Universe is flat and has no curvature. In this case the Universe expands for ever, but asymptotically reaches a steady state. When Ω > 1, the Universe is closed, that is to say, it initially expands and then starts contracting, finally ending at a point. For decades, the big question was: "What is the value of Ω?" At last an answer seems to be emerging but that is discussed in a later chapter.*

BOX 6.3

Georges Le Maître was born in Belgium in 1894. After attending a Jesuit school, he entered the Engineering College of the University of Louvain in 1911. When the First World War broke out in 1914, Le Maître joined the Belgian Army and was decorated for bravery. In 1918, after the war was over, he resumed his university studies but switched from engineering to mathematics and physics. Simultaneously, he also enrolled in courses in philosophy. His ambition was to specialise in physics and metaphysics!

In 1923, Le Maître wrote a thesis on relativity and gravitation and this won for him a scholarship from the Belgian Government, enabling him to travel to Cambridge. There he came in contact with Eddington (see ***Chandra***) and this spurred his interest in Cosmology on the one hand and numerical calculations on the other. From England Le Maître went to America, made the acquaintance of many astronomers there, and spent some time in the famous MIT in Boston. In October 1925 Le Maître returned to the University of Louvain where he remained for the rest of his life.

By 1920 Le Maître had become strongly interested in Einstein's Theory of General Relativity, which he mastered entirely by self-effort. By the mid-twenties, Le Maître, unaware of the work done in Russia by Friedmann, examined critically some of the earlier works of de Sitter on Cosmology and derived some new results. He published these in an obscure journal, and not surprisingly the results went unnoticed. In 1927 there was a big conference on physics held in Brussels (the capital of Belgium) attended by all the bigwigs, including Einstein. Naturally, as a Belgian, Le Maître was present and he took the opportunity to catch the attention of Einstein and tell him about his results. The Master's reaction was cold. He simply said (in French), "Your calculations are correct but your physical insight is abominable."

In January 1930 there was a meeting of the Royal Astronomical Society in London, during which there was a lot of discussion about Hubble's new discovery about the expanding Universe. de Sitter spoke about some of his recent results while Eddington remarked that he himself was working on the problem. All this was duly reported in the February issue of the journal *The Observatory*, which Le Maître read in full. Immediately, he dashed off a letter to Eddington reminding him that as early as 1927 he had sent him (Eddington) a paper that predicted an expanding Universe. Eddington then remembered that indeed he had received a copy of a paper but had completely forgotten about it. He now made amends by writing a letter to the journal *Nature* drawing attention to Le Maître's brilliant work three years before. Suddenly, Le Maître became a celebrity.

In May 1931, Le Maître published in *Nature* a paper that ventured to suggest that the Universe was born from an infinitesimal supreme state of matter condensation, the Primeval Atom as Le Maître called it. The explosion of this primeval atom was what later started off the expansion of the Universe. Rather colourfully, Le Maître referred to this explosion as a 'day without yesterday'. Le Maître's theory has also been referred to as a 'fireworks theory of the beginning'. As you can see, in some respects, Le Maître rediscovered what had already been found earlier by Friedmann, though perhaps with a bit more physics than just

Contd...

BOX 6.3

Contd...

bare Cosmology. No wonder Russians are annoyed with the credit given to Le Maître, overlooking the claims of their scientist.

Incidentally, triggered by his discussions held with American astronomers about the expanding Universe, Le Maître published a paper in which he provided room for the cosmological constant Λ, first introduced by Einstein and later disowned by him when he was convinced that the Universe did expand. In 1949, on the occasion of Einstein's seventieth birthday, Le Maître wrote an article in the commemoration volume in which he made a favourable reference to the cosmological constant invented by Einstein. But years had not mellowed Einstein with respect to this matter. In a terse paragraph of just five lines, Einstein curtly dismissed the article of Le Maître written in his honour! Le Maître passed away in 1966.

6.4 Hubble's epoch-making discovery

We now come to the year 1929 when Edwin Hubble of America (see BOX. 6.4 and BOX. 6.5) created a major sensation with his discovery that the Universe was actually expanding. I should mention here that as late as 1920, most astronomers still believed that the Milky Way was the entire Universe and that it was unchanging. It was only sometime later that people got adjusted to the idea that there were many more galaxies in the Universe!

Getting back to Hubble, he discovered that no matter from where one looked, galaxies seemed to be flying away from each other, which meant that the Universe was expanding. Hubble's discovery was not a theoretical deduction—it was an observational finding. The Universe was *actually* expanding—that is what Nature proclaimed, and that is what Hubble deduced from his experimental observations. If that is what experiments say, then that is the way it is! Figure 6.6 explains the meaning of the Universe expanding. What Hubble actually saw in his telescope was galaxies flying away from us. However, this 'flying away' was not because the galaxies were themselves moving but because the texture of spacetime *itself* was expanding. In other words, even if a galaxy that is far away remained perfectly stationary, it would still appear to be moving away from us at a tremendous speed, corresponding to the expansion rate of the Universe. This is the important point.

When Einstein heard about Hubble's discovery, he apparently cursed himself for suspecting his original field equations and introducing the cosmological constant in order to 'fix' the equations. In a rueful mood he is reported to have declared, "That was the greatest blunder I have committed in my life!"

In a letter that he wrote to Herman Weyl, Einstein admitted that Universe could expand, adding "If there is no quasi-static world, then away with the cosmological term!"

A new era of science had begun—physicists began to seriously examine both theoretically and experimentally, aspects related to the origin and evolution of the Universe, and Cosmology now became a full-fledged branch of science in its own right.

BOX 6.4

Hubble's discovery about the expansion of the Universe came from his long and sustained observations on the motion of galaxies. It all started around 1924 with Hubble first showing that the Universe consisted of something more than just the Milky Way. Today we take it for granted that the Universe consists of billions of galaxies, but even as late as the early twenties of the 20th century, it was believed that the Milky Way was the only galaxy in the Universe and that beyond it was one huge tenuous gaseous cloud. Hubble showed that there did exist another galaxy beyond the Milky Way and that was the Andromeda.

In 1914, astronomer Silpher noted that light from distant nebulae looked redder. When Hubble came to know of this, he realised that this might be because the Universe is expanding. And thus started his observations on the spectral shift of the light emitted by galaxies.

Recall that when a source emitting a signal of frequency ν is moving, a stationary observer detects the signal as having an altered frequency ν'. The frequency difference $\Delta\nu = \nu' - \nu$ is called the *Doppler shift*. Doppler shift is observed not only with sound waves (as must be known to you) but also with electromagnetic waves. Thus, if the light source is moving towards the observer, then $\nu' > \nu$, while if it is moving away, then $\nu' < \nu$. In the former case the light is blue-shifted, while in the latter case, the light has a redshift.

The spectral shift z is defined by

$$z = \frac{\nu - \nu'}{\nu'}.$$

There is an important point that must be appreciated here. Let us say we are looking at a nearby galaxy. This galaxy that we are observing may be moving towards us or away from us; and there would correspondingly be a Doppler shift. Suppose we were looking at a galaxy that is far away. That too may move towards us or away from us. But *superposed* on this movement, there would be a *general movement due to the expansion of the Universe as a whole*. You can understand this with the help of the following analogy.

Consider a balloon that is expanding. Let us say you are observing from a point A and that there are two ants at points 1 and 2 at different distances from you. If point 2 is farther away, it would recede faster than point 1. In case point 2 is sufficiently far, then the individual movement of the ant at point 2 would not matter much, and it would appear to move away to the observer.

Let us apply all this to galaxies. If observations are made of distant galaxies, then the spectral shift seen would be indicative mainly of the expansion rate of the Universe, rather than of the individual movement of the galaxy. This is the point that Hubble kept in mind while making observations.

Hubble patiently analysed the data for z measured by him and others as well, for a number of distant galaxies. Next, he wanted to know whether there was any law underlying these observations. In particular, he wanted to know if z was related to the distance of the galaxy from the observer. Hubble found that there did exist such a relation given by

Contd...

Box 6.4

Contd...

$$z = H_0(D/c),$$

where D is the distance of the galaxy, and H_0 is a constant now known as the Hubble constant. The *current* value of the Hubble constant is

$$H_0 = 100h_0 \text{ kms}^{-1} \text{ Mpc}^{-1}, \quad 1 \geq h_0 \geq 0.5.$$

The parameter h_0 is indicative of the uncertainty surrounding the value of the Hubble constant as it is known today.

Now what is so great about this discovery? To appreciate this, I invite your attention once again to Fig. 6.6. Here we find that when the balloon is inflated, the distance between *any two* dots increases—it does not matter which two dots we select. Applied to the Universe, this means that no matter from which galaxy an observer sees, he/she would find that all the other galaxies around him are moving away from him. This is precisely what Hubble found—remember we do not have any privileged position in the Universe. Thus, Hubble's discovery is a clear signal that our Universe is expanding, at least right now.

Box 6.5

Edwin P Hubble was born in Missouri, America, in 1889. He passed the Matriculation from the University of Chicago and later received the B.S degree from the same university in 1910. He was training to become a lawyer, but, through friends, developed a strong interest in astronomy. Nevertheless, after graduation he left for England as a Rhodes Scholar, to study jurisprudence in Oxford.

At Oxford, Hubble gave a good demonstration of his athletic and sporting abilities. He not only became an Oxford Blue in track events, but also took to heavyweight boxing and even defeated the French champion Georges Carpentier in an exhibition match! He was given an offer to turn professional but refused.

On return to America in 1913, he was admitted to the Bar in Kentucky. For a while he practiced as a lawyer. But his interest in astronomy prevailed, and he was preparing to return to Chicago for graduate studies in astronomy. Meanwhile, the First World War intervened, and Hubble enlisted in the U.S Army. He saw action in France, and rose to the rank of a Major.

After the war was over, he returned to pursue his interest in astronomy and thus began a most fruitful career. Hubble was a very keen observer and made many important discoveries. In 1924 he announced the discovery of stars known as Cepheids, whose intensity pulsates. Hubble cleverly used the information on the variation of the luminosity to estimate the distance of the galaxy in which these objects were located. Measuring the distance of far-away galaxies is very crucial.

Some years later when sufficient data on galactic distances had been collected, Hubble decided to examine the distance data against the redshift data accumulated by him with the data collected earlier by astronomer Slipher. When Hubble

Contd...

BOX 6.5

Contd...

did this, he found that the redshift was linearly proportional to the distance. This meant that distant galaxies (which were older) were moving away faster than the nearby galaxies, giving some clue to expansion rate of the Universe. About all this, Hubble said,

> "(all this) should furnish a clue as to the exact nature of the Universe. It may then be possible ..…. to say if the Universe and space itself is expanding at a rapid rate and in a remarkable manner. And finally, it may be possible to describe the nature of the expansion and to determine the time at which the expansion began—that is to say, the age of the Universe."

The Second World War interrupted Hubble's research. He tried to enlist in the infantry as he did during the First World War, but was prevailed upon to become the Chief of Ballistics and the Director of the Wind Tunnel Laboratory in the Aberdeen Proving Grounds in Maryland. Hubble served in this capacity till the end of the war. He died in September 1953.

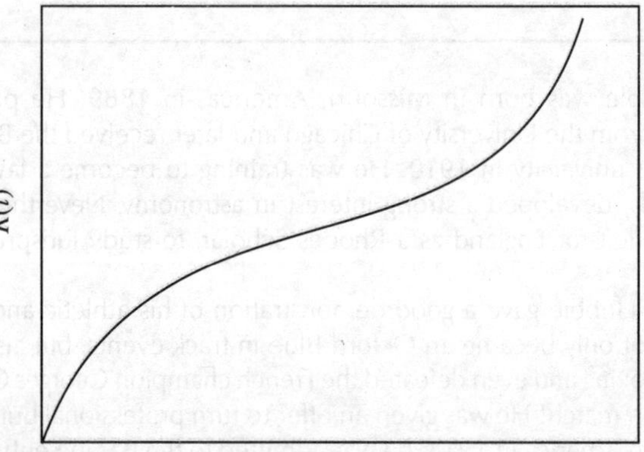

Fig. 6.5 *This figure shows schematically the essence of Le Maître's model. As can be seen, qualitatively it is similar to one of the models of Friedmann discussed in Fig. 6.4.*

6.5 Gamow and his Big Bang

Moving on, we now come to the year 1948 when Gamow [see BOX 6.6], along with Ralph Alpher and Hans Bethe, took the ideas of Le Maître further. Gamow was basically concerned with the origin and evolution of the elements from the original homogenous state of extremely high density and high temperature that is implied by the Le Maître model. He said that "the evolutionary development of the Universe must have started billions of years ago from a homogeneous state of high density and temperature". In short, Gamow speculated that at the start matter in the Universe was essentially a over-heated neutron fluid. When the expansion of the Universe began, the density began to drop and neutrons started decaying into protons. Soon, neutrons started combining with protons to form nuclei, and Gamow argued that this was how the elements started forming. From this Gamow went on to estimate the masses and the sizes of the galaxies.

Based on Gamow's guess about the temperatures prevailing at the time the light elements came into existence, his former colleague Alpher, in association with Herman, later deduced that if the Universe was very hot at that time, then remnants of that primordial radiation must still be filling the Universe, and currently have a temperature of about 5 K. This was around 1948–1950, and at that time no one ever thought that it would be possible to observe the relic of the original fireball. Today, we invariably use the word *Big Bang* to describe the birth of the universe in a primordial fireball, but Gamow did not use such a term. The term Big Bang was introduced by the British Astronomer Fred Hoyle [see BOX. 6.7] in his broadcasts over the BBC; it has subsequently become a part of physics! In passing I might call attention to how Gamow sought the co-authorship of Bethe so that the paper would have three authors [Alpher, Bethe, Gamow] whose names sounded like the Greek alphabets α, β, γ!

6.6 Discovery of the Cosmic Microwave Background

Cut now to the early sixties. In 1964, Robert Dicke in Princeton University constructed the apparatus needed for the search of the remnants of the primordial radiation, but even as he was getting ready, two other people who were not at all looking for this cosmic background, literally stumbled on it by chance! More now of that exciting drama.

In the early sixties, America was making an all-out effort to introduce satellite communications, and in this context had launched a satellite called Telstar. Two scientists, Arno Penzias and Robert Wilson of the Holmdel division of the famous Bell Laboratories, were involved in receiving microwave signals from the Telstar. [See also BOX 6.8 and BOX 6.9.] For this purpose they used a huge 7 metre horn antenna. The antenna was very sensitive but it had been carefully built as to avoid picking up radiation emitted by the earth. Thus, when the antenna was pointed towards the sky, no signal was expected excepting of course from the Telstar. Penzias and Wilson directed the antenna towards the sky when Telstar was not in sight; yet it picked up a signal whose origin was mysterious. Initially, Penzias and Wilson believed that this was spurious noise. They then did everything possible to get rid of this 'noise'. They cooled the detector. They drove away two pigeons which had parked themselves inside, and they also thoroughly cleaned the contamination due to bird droppings. They further covered the joints with metallic tape, and so on. But try as they might the signal that they kept picking up from the sky simply refused to go away. And what was even more interesting, they picked up the same strength of signal no matter in which direction they pointed the antenna. Was it really some microwave signal coming from out there in the space? If so, where did it come from? If it was from some stellar object, it ought to

BOX 6.6

George Gamow was born in Odessa, Russia in 1904. Even as a boy, Gamow became very much interested in science. He began his university career in Odessa but was transferred a year later to Leningrad (now, back to its old name of St. Petersburg!). Meanwhile, the Bolshevik revolution of 1917 had taken place, and Russia now became a part of USSR.

In 1928, Gamow had a chance to go out of Russia to Gottingen in Germany. In those days, Gottingen was one of the world centres for physics. There he made his first great discovery—the quantum mechanical tunnelling of the alpha particle as the mechanism responsible for the alpha decay of heavy elements like uranium. This discovery made Gamow a hero in Russia.

He returned to his native land in 1931 but found that conditions had changed drastically; Russia had now become a tightly-controlled dictatorship. Gamow found the atmosphere too stifling. He and his wife then planned an escape. They purchased a small canoe and hoarded food for months. They then managed to wrangle permission for having a holiday on a Black sea port. And one day, they tried to row out of Russia—they had to paddle nearly 250 km and they thought they would be able to do it. But it did not work that way; the canoe was caught in a storm and blown back to the shore, forty eight hours later! The chance to leave Russia came two years later when the Government sent him to Brussels to attend a conference. Gamow went and never returned!

Gamow went to America in 1934, and he stayed there for the rest of his life. During the Second World War, Gamow worked on the atom bomb project. After the war he turned to the origin of elements in the early Universe.

In 1953, a friend walked into his office and told him about the discovery of the structure of the DNA. As you know, the DNA has a double helix structure. The question was: How is the information in the DNA decoded for synthesising proteins? Gamow immediately became fascinated by the problem and came up with an ingenious solution to the genetic code problem. However, it later turned out that Nature does not quite work that way. Nevertheless, Gamow's work on the genetic code highlights his problem-solving approach to science. He would pick up problem that sounded like a good puzzle, and then try to crack it. First it was in nuclear physics, then in astrophysics, and finally in molecular biology.

This man who flirted with such diverse subjects was a colourful personality in many ways. He was exuberant, jovial, enjoyed entertaining people with magic and jokes. He could draw cartoons and made a fine art of explaining complicated physics in popular terms. He was one of the first great popular science writers, and his Mr. Tompkins books inspired me much. No wonder he won the Kalinga Prize of the UNESCO for the popularisation of science.

I saw Gamow in 1959 in Bombay. He was then lecturing at the Tata Institute, where I was working at that time. Niels Bohr had also come to Bombay then, and one day there was a garden party in the honour of Bohr. On that occasion, Gamow downed no less than seventeen cups of cold coffee. It was said that Gamow who was a heavy drinker, found the Prohibition in Bombay unbearable and resorted to coffee drinking as a substitute! Shortly before he died in 1968, Gamow said, "Finally, my liver is presenting its bill!" a reference to his drinking. He combined physics with magic, a carefree attitude, sport, fun, and jolly good living!

BOX 6.7

Sir Fred Hoyle was one of the most original astrophysicists of the second half of the twentieth century. Born in 1915 in Yorkshire in England, Hoyle developed a liking for science in general and astronomy in particular, at an early age. After school he went to Cambridge, passed the Mathematics Tripos winning the Mayhew Prize, and followed this by winning the Smith Prize. He was then elected a Fellow of St. Johns College which fetched him a salary. But now he had to pay income tax; however, he discovered that there was a way of avoiding tax. This is what Hoyle himself says about it.

> "But I did discover that when I began to get a salary, as long as I could claim to be a student, I was free of income tax. That is why Dirac agreed to be my research advisor. He thought it amusing. He did not want students, but here was a student who did not want a research director. That was typical of Dirac's sense of humour."

Hoyle started his research career as a physicist by working with M.H.L. Pryce on weak interactions. However, his friendship with astronomer Lyttleton diverted his interest to astronomy. Soon, Hoyle was applying nuclear physics to stellar problems. One of his major contributions was describing how the elements came into existence.

After the Second World War, Enrico Fermi and several others began to speculate on how the first elements came into existence in the Universe. George Gamow in particular, as you know, gave much thought to this problem. They all concluded that element creation must have taken place when there was a high flux of neutrons. Everyone decided that this must have occured in the very early stages of the Universe when the density was high, and there were plenty of neutrons as well. However, using this scenario one could only speculate on how some of the lighter elements, up to ^{7}Li, could be built. Some other setting was needed to go beyond lithium. Hoyle conjectured and later showed through extensive calculations performed with William Fowler of America, how elements up to iron could be cooked in succession in the cosmic cauldron called the star. I have referred to this in the volume ***Chandra.*** Not only that; Hoyle also showed that elements beyond iron must have been prepared during supernova explosions. Considering the detailed and consistent accounts given by Hoyle regarding the evolution of elements, it was a great surprise when he was not awarded the Nobel Prize along with Fowler for the theory of the origin of elements.

Many would perhaps remember Hoyle for proposing the so-called Steady-State Theory of the Universe, which he offered as an alternate to the Big Bang model. The origin of this theory has an interesting history. Once Hoyle gave a seminar in Birmingham in the presence of Sir Rudolph Peierls. Hoyle was describing how the elements were prepared, one by one, starting from hydrogen of course. Peierls then asked, "Where does the hydrogen come from?" Hoyle did not give a reply then; but the question remained with him. A film that he later saw in the company of Gold and Bondi gave him the idea that possibly, hydrogen was being constantly created in the Universe, even as it was expanding. Hoyle supposed that Nature was creating enough hydrogen to keep the density of the Universe a constant, even as the Universe was expanding. Interestingly, Hoyle's movie companions

Contd...

BOX 6.7

Contd...

Bondi and Gold independently got the same idea and they too proposed a steady-state theory of the Universe. Hoyle's model was, however, more detailed.

Shortly after this came Gamow's model and Hoyle in his radio broadcasts over the BBC humorously dismissed it as the Big Bang model. Thus, the name Big Bang is really due to Hoyle.

Right from the beginning, the Steady-State Theory of Hoyle met with opposition. In fact, this opposition started even before the discovery of the cosmic microwave background radiation. One of the strongest critics was fellow astronomer Martin Ryle of Cambridge. Based on the strength of his radio astronomy observations (relating to the number of radio sources in the Universe), Ryle bitterly attacked Hoyle, more than his data warranted. Later when Penzias and Wilson made their discovery, Hoyle contributed many clarifications to certain aspects of the Big Bang model, but he never reconciled himself to the idea that the Universe had a unique birth. Till the end he stuck to the view that the Universe always existed, though it was growing at the present time.

Hoyle was appointed the Plumian Professor in Cambridge, a very prestigious post. Earlier, during Eddington's days, Rutherford made sure that nuclear physics was given more importance than astronomy. Now in the post war years, the same battle continued—astronomy versus the rest. After much agitation Hoyle managed to found the Institute of Theoretical Astronomy that later produced celebrities like Martin Rees and Stephen Hawking.

Hoyle's achievements went much beyond astronomy. He wrote science fiction novels, produced popular television serials and also broadcast a series of talks entitled *The Nature of the Universe* over the BBC. Hoyle entered radio broadcasting by chance. It so happened that one person who was to have done a series could not do so, and an urgent replacement was needed. He was asked whether he was ready to take over. When Hoyle learnt that he would be paid two hundred and fifty pounds for giving just five talks, he jumped at the offer because his annual salary at that time was only four hundred pounds!

Towards the end of his life, Hoyle became interested in the origin of life. Once again he proposed a theory quite at variance with the orthodox view. The widely accepted picture is that life emerged here on earth roughly one billion years after it was born. It emerged as a primitive living molecule in what is called the pre-biotic soup, which is the name given to the oceans of those days. Hoyle instead suggested that life on earth developed from living molecules that were transported from outside by comets.

Despite his unorthodox views on many matters, Hoyle commanded much respect and won many honours. He died in 2001.

disappear when the antenna was turned, but this signal never vanished. What did it mean? Penzias and Wilson had no answers, only an inexplicable microwave signal.

One day, during that period, Penzias happened to make a phone call to Professor Bernard Burke of MIT. In the course of their conversation Penzias casually mentioned the mysterious signal he and Wilson were picking up. For Burke, this remark rang a bell. He had sometime earlier heard from Ken Turner of

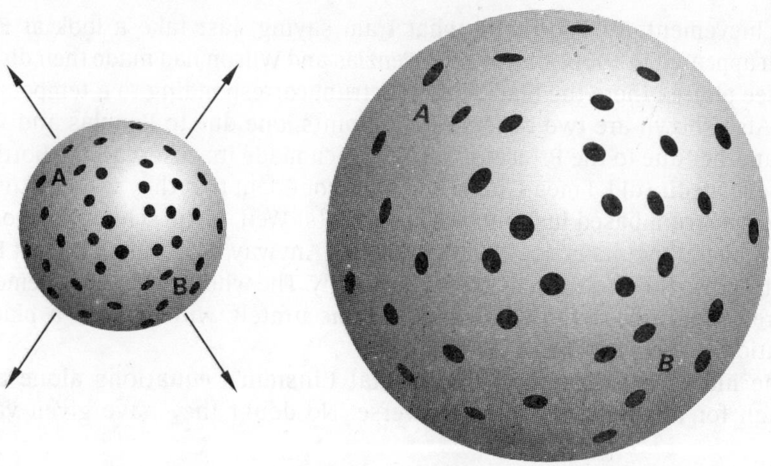

Fig. 6.6 *This figure is the popular way of explaining Hubble's discovery. Imagine that the Universe is like the balloon illustrated above, the dots on it representing the galaxies. When the balloon expands, the dots move away from each other. An observer may sit on any dot and look around and he would see all the other dots flying away from him on account of the expansion of the balloon. This is what Hubble essentially found, with respect to the Universe.*

Carnegie Institute that Peebles of Princeton University was saying that "there ought to be a background of radio noise left over from the early Universe," and that Professor Dicke of Princeton was in fact searching for this background radiation. Penzias then phoned Dicke, who sent him a copy of the paper by Peebles in which this background radiation was clearly mentioned. Very soon it became clear that the 'unwanted signal' that Penzias and Wislon were picking up was in fact the remnant of the radiation that originated at the time of the Big Bang! Roll and Wilkinson of Dicke's group were *actually* trying to look for this *cosmic microwave background* (CMB) radiation. They did not find it. Instead, two people who did not even know that such a cosmic relic existed, stumbled upon it. Well, that is luck! However, you must concede this: Penzias and Wilson did meticulous tests before becoming convinced that something new was there. That is the way good science is done. Shortly after this, Penzias and Wilson announced their discovery in a short paper in the *Astrophysics Journal,* and back to back with their paper there appeared the paper by Dicke and co-workers. In their paper, Penzias and Wilson say, "A possible explanation for the observed excess noise temperature is the one given by Dicke, Peebles, Roll and Wilkinson in a companion letter in this journal." Penzias and Wilson detected the CMB and Dicke explained to the world that the signal observed by Penzias and Wilson was *indeed* the CMB. In 1978, Penzias and Wilson received the Nobel Prize for their lucky and chance discovery.

There is one important thing I would like you to appreciate at this point and that is, Penzias and Wilson observed the CMB at a single frequency—4080 Mhz or 7.5 cm. Now blackbody radiation has an entire spectrum, and to talk of the temperature of the radiation from a measurement at a single frequency is quite an achievement. To appreciate what I am saying, just take a look at Fig. 6.7 which appeared in 1966, shortly after Penzias and Wilson had made their discovery. You see plotted there the blackbody spectrum corresponding to a temperature of 3 K. Also shown are two observational points, one due to Penzias and Wislon, and another due to the Princeton group which made its observation shortly after the Bell Lab duo did. I mean, people were so confident that they were talking of an entire spectrum based just on two data points! Well, if the science is good, then the number of points is not always an issue! Anyway, take a look now at Fig. 6.8 which shows the spectrum as mapped recently. The whole range is extremely well covered, and indeed the shape fits very accurately with that of a blackbody radiation.

The history of the CMB tells us that Einstein's equations alone are not enough for understanding the Universe. No doubt they have given valuable

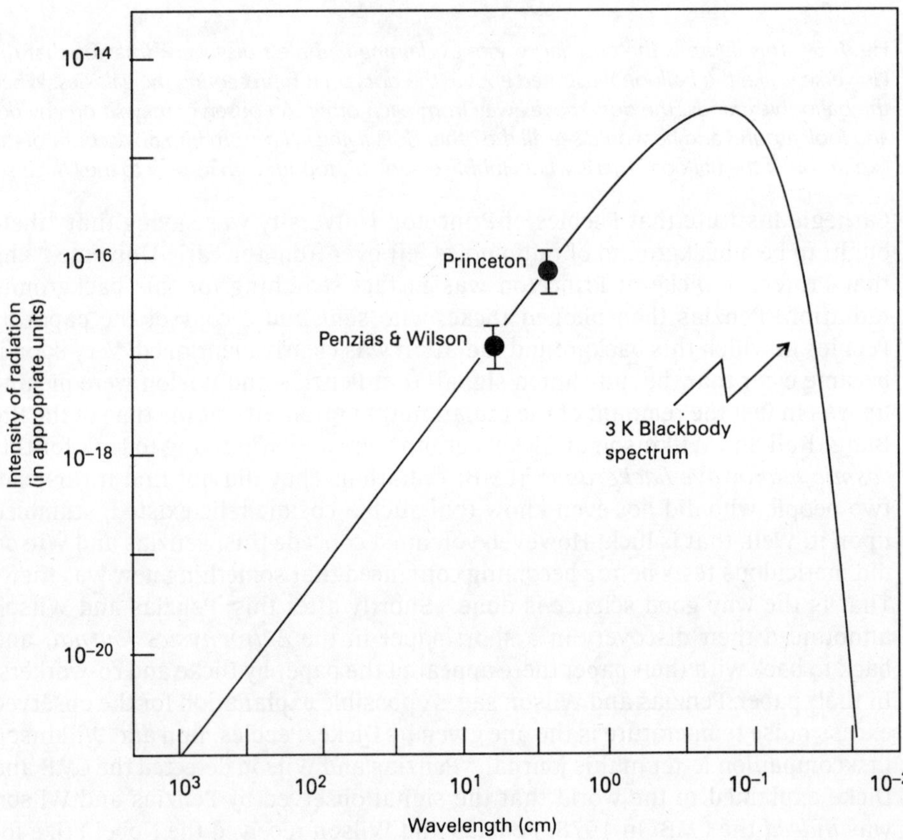

Fig. 6.7 *This figure shows the blackbody spectrum for a temperature of 3 K, and the two points that represent the very first observations. Figure 6.8 highlights the great daring with which the original authors Penzias and Wilson of Bell Labs on the one hand and the Princeton group of Dicke on the other, asserted that the microwave radiation they had observed had a blackbody spectrum.*

clues, but for a more detailed picture of the evolution of the Universe it is necessary to bring in nuclear and particle physics as well. The closer we go in evolution history to the Big Bang, the smaller is the length scale that must be considered. This automatically means that we must go from electromagnetic interaction to weak and strong interactions, and finally to gravitational interaction. Thus, one can visualise the post-Big-Bang evolution along the lines of Fig. 6.9 and Fig. 6.10.

Starting with the Big Bang, there are many eras, each epoch relating to a regime where some forces were dominant. Correspondingly, the particle species that prevailed has also varied with time. When Gamow did his work we knew very little about the weak and strong interactions. Therefore, he could not reconstruct the scenario at times smaller than about say 100 seconds after birth. Some years later, when Weinberg wrote his famous best-seller, *The First Three Minutes,* the limit could be pushed to about 10^{-12} seconds after birth. Today, we could perhaps do a bit better; but we still have a long, long way to go before we can talk confidently about what happened, say at one Planck time after the birth.

Let us go over Figs. 6.9 and 6.10. These figures must be seen in conjunction with the splitting of the four forces. What we notice then is that as diversity progressively evolves from the initial homogeneity, the Universe becomes more and more complex. In all this there is one basic point to be noted. Initially

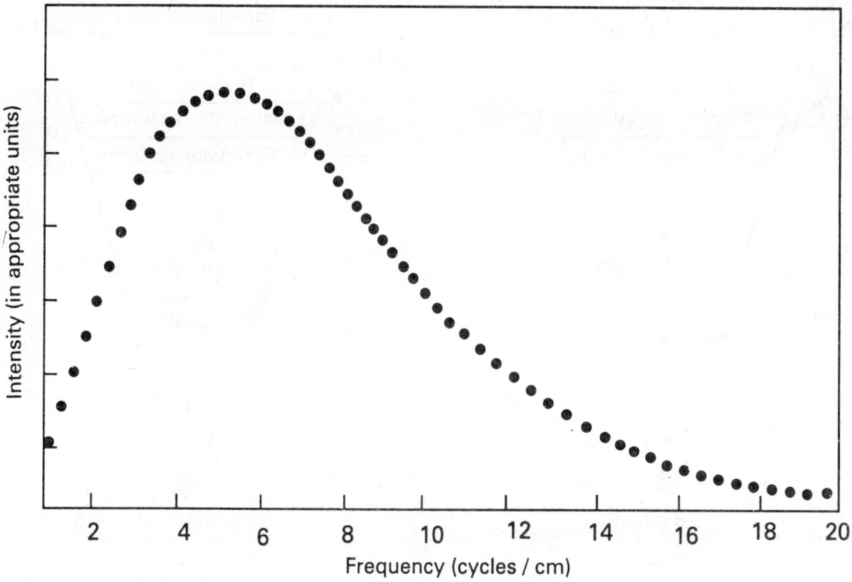

Fig. 6.8 *This figure shows the current status of the spectrum of the cosmic microwave background obtained using the satellite COBE (Cosmic Microwave Background Explorer). The unit 'cycles/centimetre' is obtained by dividing frequency by the velocity of light. The observed spectrum fits the blackbody spectrum beautifully, and the present value of the temperature is 2.728 K.*

when the size of the Universe was small and the density was high, the photons collided frequently with particles. Later, after expansion, there came a stage when photon collisions became less frequent; thereafter matter and radiation have coexisted almost as if they are independent entities. This is usually referred to as the decoupling of radiation from matter.

Let us now look at this decoupling in slightly greater detail. As I just remarked, when the temperature is high, photons scatter frequently after colliding with particles of matter. Let us now come to the regime when the temperature is relatively low and it is the scattering by electrons that is important. Where electrons are concerned, photons can scatter off electrons via Compton scattering process and Thomson scattering process. Compton scattering ceases to be important below 0.5 MeV or 10^7 K and Thomson scattering below about 14 eV or about 10^5 K. By the time the temperature comes down to say 10,000 K, most of the electrons would have already combined with nuclei to form neutral atoms, thus drastically reducing the supply of free electrons. Thus, everything taken together, when the

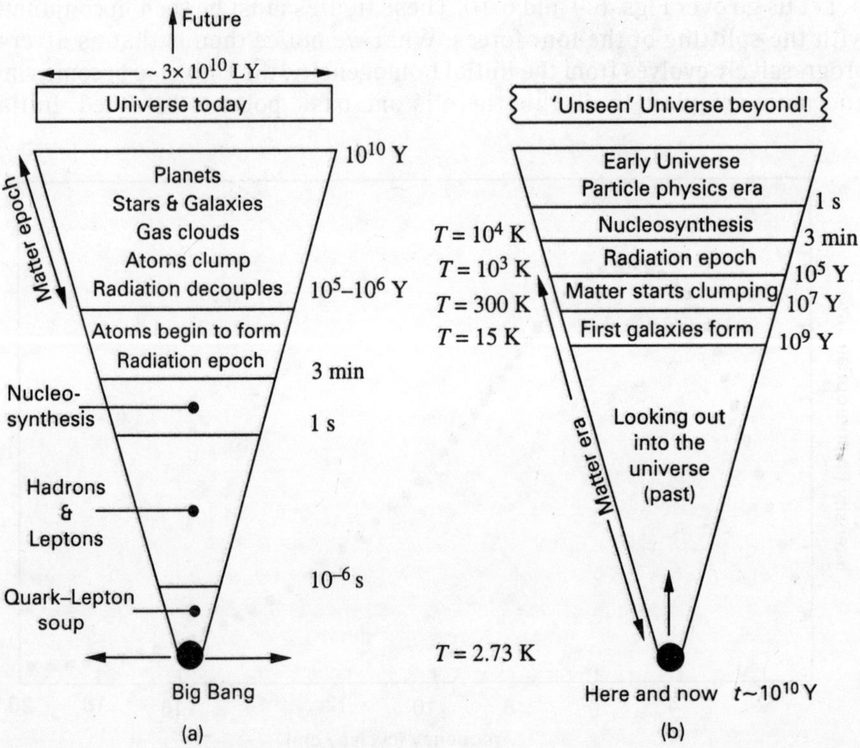

Fig. 6.9 *Shown here is the history of the Universe in two different ways. In (a) we have the cosmic picture with Big Bang at the bottom and the present at the top, while (b) is a rendering of the scenario when we look out into space and thus back into time. Caution! In this case the lateral extent of bands DO NOT represent the size of the Universe. Actually, as we look into the past we would be looking at a smaller size. However, the portion we are looking at might well be a part of a larger Universe that remains out of sight always!*

Universe cools down to about 10,000 K, photons can travel long distances without bouncing off electrons. This is when the Universe is said to become *optically thin*, and this is also when radiation is said to decouple from matter – current estimates say that this happened roughly somewhere between 100,000 to 300,000 years after the Big Bang. Note incidentally that if the Universe were not optically thin today, we would not be able to do any astronomy! The light coming from a distant star would get scattered away long before it reached the earth!!

Getting back to Fig. 6.10, as can be seen, there have been many epochs since the birth of the Universe. This separation into distinct epochs is only a rough guess. We still do not know many things, particularly because observational data concerning many aspects of the Universe is lacking. Even so, the picture is good enough for a start, and we also have a good idea of many of the questions that need to be resolved. I cannot list all of them, and will mention just a few. Answers to some have since been found while the answers to the others are still awaited.

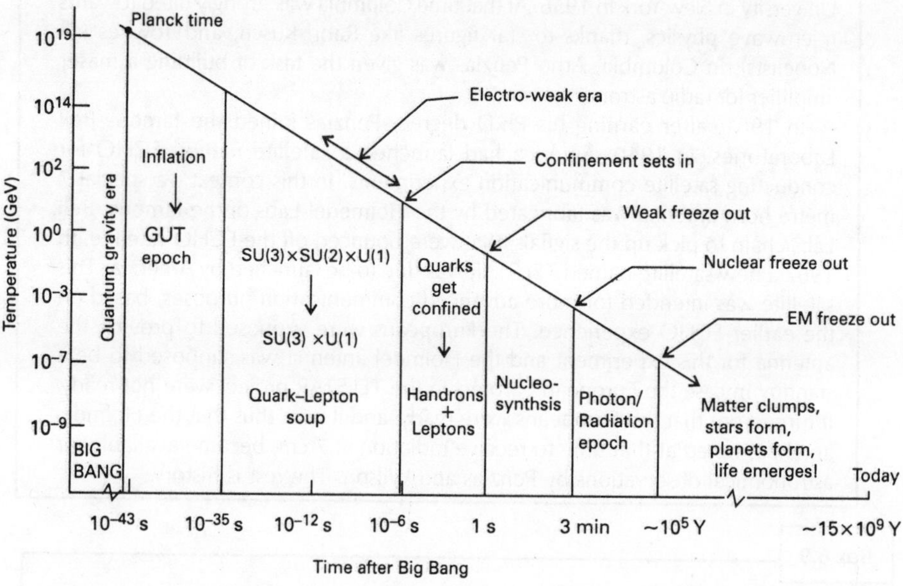

Fig. 6.10 *This figure presents the information about the various epochs of the Universe in a different manner, and with some more details concerning the early phase. It has to be studied along with the sketches in the previous figure. (CAUTION: The axes are highly nonlinear).*

6.7 Chapter summary and the road ahead

In this introductory chapter on Cosmology, we have noted the following.

- It all started when Einstein applied his theory of gravitation to the entire Universe.
- Einstein discovered to his horror that the Universe was not static as he believed

Box 6.8

Arno Pernzias was born in Munich, Germany, in 1933. Towards the end of the thirties, the Nazi hatred for Jews began to take a particularly virulent turn and Jews were being rounded up and sent to various concentration camps. Penzias' family was supposed to be deported to Poland. The father decided that the family should escape from Germany and head for America. Arno and his brother were both put on a train bound for England; each had a suitcase with their initials painted on it. Their mother managed to go England independently and join them. The father came soon after. The entire family landed in New York in January 1940.

Life was not easy for them but the fear of persecution was no longer there. Arno went through school and entered the City College of New York. That was when he 'discovered' physics as he puts it, making the switch from chemical engineering. After a small stint in the U.S. Army Signal Corps, he entered Columbia University in New York in 1956. At that time Columbia was strongly tilted towards microwave physics, thanks to star figures like Rabi, Kusch, and Townes (all Nobelists). In Columbia, Arno Penzias was given the task of building a maser amplifier for radio astronomy work.

In 1961, after earning his Ph.D degree, Penzias joined the famous Bell Laboratories. In 1960, America had launched a satellite named ECHO for conducting satellite communication experiments. In this context , a special 7 metre horn antenna was fabricated by the Holmedel Labs of the famous Bell Labs chain to pick up the signals that were bounced off the ECHO satellite. In 1962 a new satellite named TELSTAR was due to be launched by America. This satellite was intended for more advanced communication purposes, based on the earlier ECHO experience. The Europeans were supposed to provide the antenna for this experiment and the Holmdel antenna was supposed to be a standby in case the European partners in the TELSTAR project were not ready. It turned out that the Europeans were ready, and it was thus that the Holmdel antenna, tuned at that time to receive radiation at 7 cm, became available for astronomical observations by Penzias and Wilson. The rest is history.

Box 6.9

Compared to Arno Penzias, Robert Wilson had a smooth and traditional American childhood. He was born in Texas in 1936 and attended Public School in Houston where, as he puts it, his 'career was undistinguished except for math and science.' Both his parents were inveterate 'do-it-yourselfers', almost no task being beneath their dignity or beyond their ingenuity. The son was duly infected, and he too picked up various tinkering skills from auto and electronics shops. As he records, 'I used to fix radios and later television sets for fun and for spending money. I built my own hi-fi set and enjoyed helping friends with amateur radio transmitters, but lost interest as soon as they worked.'

From School, Wilson went to Rice University for his B.A degree. Next stop was Caltech or California Institute of Technology to give its full name, made famous by many great names (for example, Feynman and Gell Mann). He was not quite

Contd...

Box 6.9

Contd...

sure what he was to going to do by way of his thesis project, when a visiting astronomer from Cambridge University in England suggested radio astronomy as a field for study. Wilson liked the idea because radio astronomy offered a nice mix of physics and electronics.

At that time, Caltech was setting up what was known as the Owens Valley Radio Observatory, and Wilson was supposed to do his thesis project there. His thesis project was to be interferometry based on hydrogen emission lines but the required maser amplifiers did not work. So he had to change his line of investigation to complete his thesis requirement. Later, the interferometer was built with help from Bell Labs.

After this experience Wilson joined the Bell Labs at Crawford Hill in 1963. The 7 metre horn-reflector antenna, originally built for satellite work, was sitting there waiting to be used and Wilson and Penzias started preparing it for use in radio astronomy. At that time there were already much larger telescopes available; so what was the big deal about a small horn antenna? Wilson, with his experience in measuring antenna temperatures, quickly realised that this antenna had two very good plus points. First, it picked very little of the radiation that is reflected by the Earth. Next, it could be accurately calibrated by pointing to a transmitter that was just 1 km away. Thus, in between various plans for radio astronomical observations, Wilson, along with Penzias, started on an accurate study of the temperature of their antenna. By this I mean essentially its noise level.

You see, every instrument designed to detect a signal has to contend with unwanted signals called noise. The noise becomes very important particularly when one wants to detect very weak signals. The lower the noise, the more sensitive the detector. Noise in radio work has a spectrum whose signature is temperature. The lower the noise temperature, the better the detection system and the greater is its sensitivity. Wilson along with Penzias was engaged in measuring the noise temperature of the Holmdel antenna and trying to bring it down as much as possible.

It was in the course of these extensive studies that Penzias and Wilson detected the unambiguous presence of excess radiation with a temperature of about 3 K. The rest of the story is told in the main body of the chapter.

it was. To fix this 'defect' Einstein introduced the famous cosmological constant.

- Freidmann and Le Maître who came after Einstein found that Einstein's equations were correct and that Einstein was wrong! The Universe had to have a birth and then expand.
- Meanwhile, Hubble discovered that the Universe does expand.
- After this it was the turn of Gamow to give a physical picture of the infant Universe. It was essentially a tiny fireball, the cauldron in which the very first elements were cooked.
- Implicit in Gamow's picture was a background electromagnetic radiation.
- Penzias and Wilson discovered the remnants of that radiation, confirming that the Universe did originate in a Big Bang.

The following chapters would deal with many of the issues and problems that arose after these initial discoveries. For the present let me just mention, without explaining in detail, a few of the issues that came up in the years following the discovery of the CMB. They are:

- How come the Universe is so homogeneous?
- Even though the Universe appears quite homogeneous on large scale, there are inhomogeneities on a small scale. They must have been there right from the early times – otherwise galaxies could not have formed. How did these initial inhomogeneites come about?
- Why is there more matter than anti-matter in the Universe?
- Is there any non-luminous or dark matter in the Universe? If so how much and what is its nature?
- Are there any monopoles?
- If the Universe did originate in a Big Bang, does it mean that the density of the Universe was infinite at that point of time? In other words, did the Universe originate in a singularity?

These are some of the questions that would come up in the chapters to follow.

7

MAKING THE TWO ENDS MEET—A CURTAIN RAISER

The title of this chapter ought to give you an indication of what I shall be attempting presently. In Volume I, I focussed entirely on elementary particle physics, bringing you up to the Standard Model. In the present volume I started off with some glimpses of particle physics lying beyond the SM. And in the last chapter I gave you a brief introduction to Cosmology, starting with Einstein's investigations based on GR and then moving on to the study of the evolution of the Universe based on nuclear and particle physics. From now on, I shall try to bring particle physics and Cosmology together, and this chapter is in the nature of a curtain raiser to this process.

This coming together of particle physicists and cosmologists is an extra-ordinary happening. In a manner of speaking, the particle physicist is looking into a 'microscope' that gives him a peep into the microcosm. The cosmologist, on the other hand, seeks the help of his astrophysicist friends who all the time look through telescopes into the farthest reaches of the cosmos. The cosmologist cultivates friendship with the astrophysicist because the latter is interested in the macrocosm. (See also Fig. 7.1.) But God in His infinite wisdom has created the Universe in such a manner that the microcosm is in the macrocosm and vice versa. (You might note in passing that this is true of living systems also – the tree is in the seed, and the seed is in the tree!) No wonder then that for a complete understanding of the Universe, we need a knowledge of *both* the micro- as well as the macrocosm. This is the extraordinary revolution in scientific thought that is currently taking place—the coming together of people interested in the two 'extremes' of the Universe, if I might put it that way. And yet, there really ought to be no surprise in this because philosophers, particularly our ancients, have always felt that the macrocosm is in the microcosm and vice versa. There are thus really no two ends. As Sheldon Glashow sees it, it is like a snake swallowing its own tail, as illustrated on the cover of this volume.

From now on, I shall, as convenient, be flipping between the two topics, particle physics beyond the SM the on the one hand and cosmology on the other, telling you a bit about this and then a bit about that. You know this sort of thing happens regularly in novels, where for a while you follow some events and the fate of characters associated with those events and then switch over to other characters and what they do. This is not a novel, but as you read on, you will, I am sure, find that indeed truth is stranger than fiction! So I hope you won't find my flipping back and forth irritating. One important difference compared to a novel is that here we do not know the end of the story, if there is one! The end, if at all there is one, is in the future and I don't think it will happen in my time.

OK, so what are the topics I shall be covering from now on? Here is a short list. I have to tell you about some of the particles produced at the time when the Universe was young. I have to tell you something about why there is so much more matter in our Universe than anti-matter, the so-called *asymmetry problem.* I have to tell you about why the Universe appears homogeneous, and in that process may be you will get some idea of 'what it was that banged'. I have to tell you something about the exciting things that have happened in neutrino physics and their implications. I have also got to describe how particle physicists have gone on to doing physics not in four dimensions but in 10 and 11 dimensions! This raises the problem: We experience only four dimensions in all. Where are these other dimensions that particle physicists glibly talk about? You will find how clever the particle physicists are in hiding these extra dimensions from our view! Then there is the whole business of the ultimate fate of the Universe, and when we start discussing that, we will find that particle physicists have made our wonderful Universe into a tiny expanding bubble in a 'sea of interconnected bubbles', where each bubble represents a Universe of its own!

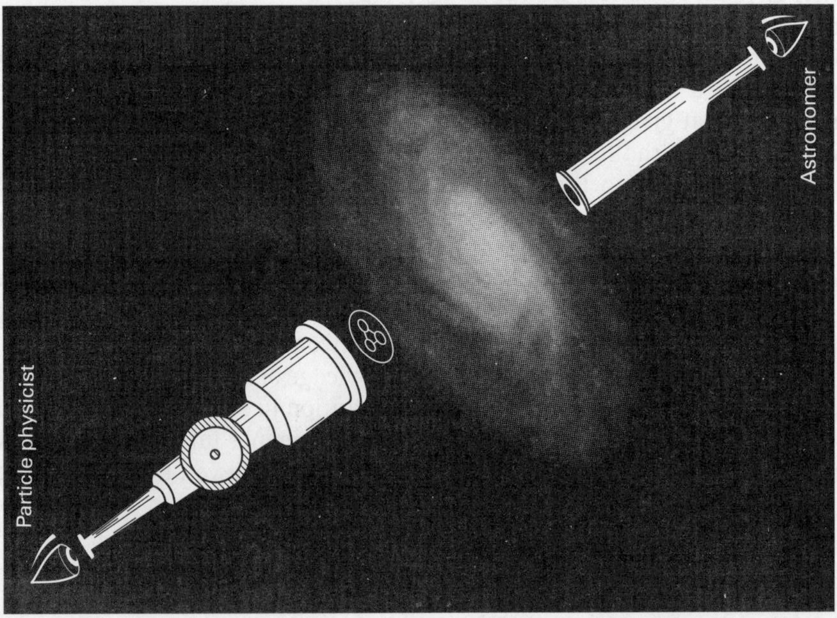

Fig. 7.1 *This figure illustrates schematically the interconnection between the physics of the microcosm/small with the physics of the macrocosm/large. The two types of physics converge because when an astronomer peers deep into space, he is really looking at the very early Universe when the Universe was tiny. And the physics that operated then was the physics of the small that particle physicists specialise in. It is only about the seventies of the twentieth century when astronomers started peering deep enough and particle physicists started looking at really small distances that this convergence emerged.*

I mean there are so many ideas—do we call them wonderful or crazy? I shall not describe them either way. I shall simply let you have a glimpse of the extraordinary imagination and creativity that has now entered into the game. Once upon a time, progress came through cute experiments, daring ideas, solid mathematics and things like that. Now, the ball game is different. It is mega experiments, sometimes absolutely new mathematics, and ideas that one would describe as bordering on the atrocious were they not from serious physicists. By the way, so fierce is the competition that no one can get away with talking nonsense! More than anything else, the current developments in this last frontier of physics highlight the extraordinarily creative, analytical and logical capabilities of the human mind. Creativity in art, music and literature might have seen great peaks in earlier eras. In physics, the superpeaks are in view for the first time. What we are seeing makes even towering achievements like relativity and quantum mechanics look like modest hills! At least, that is the feeling I get when I survey the current scene.

So, get set for this exiting finale!

8

WHY IS NATURE PARTIAL?

You and I are made of atoms, that is to say, protons, neutrons and electrons, as opposed to anti-atoms. What about the rest of the Universe? Is all of it made up of just matter, or are there huge segments made up of anti-matter? If so, which is more dominant, matter or anti-matter? These questions have been examined, and it is found that matter dominates. Why? That is the point discussed in the present chapter.

8.1 Is there matter–anti-matter symmetry in the Universe or not?

We now begin the last part of our journey, trying to get a glimpse of the intimate connection between particle physics and the nature as well as the evolution of the Universe.

From the time Dirac first introduced the idea of the positron (see *QR I*), particle physicists have accepted as an axiom that for every particle there must be an anti-particle. Yet, in this world everything seems to be made almost entirely of what we commonly call particles rather than anti-particles. Have you ever wondered about that? If anti-particles were common, we would have known about them long before Dirac told us to expect them.

In a sense it is good that anti-particles are scarce. I mean suppose you had a friend who is made of anti-atoms, that is anti-matter. He comes to meet you and you greet him with a hello. If you follow this up with a handshake, that would be the end of both you and your friend! So we should be grateful to Providence for making anti-matter scarce here on Earth.

Not only the Earth but the Moon too is made up of matter rather than anti-matter. We know this because astronauts have gone to the Moon and returned safely. It is the same with Mars also because a space vehicle has successfully landed on Mars, and we have all seen the wonderful pictures transmitted from there. In fact, the whole of the solar system is made up of only matter, and that holds for the Sun too. How can we be sure of the Sun? Well, we have caught neutrinos from the Sun, and that is what tells us that the Sun is made of matter rather than anti-matter. If it was made of anti-matter we would be receiving mostly anti-neutrinos from it. Moving next to our galaxy, the Milky Way, we can assert that that too is up made of matter. How can we say that? Thanks to the cosmic rays. There is now plenty of evidence to show that the cosmic rays received by the Earth consist predominantly of particles rather than anti-particles. Thus, up to our galaxy, we can be sure that matter very much dominates over anti-matter.

What about galaxies that lie beyond? From these we receive mostly light and since the photon is its own anti-particle, we cannot say whether the distant galaxy is made up of matter or anti-matter. But if we could detect the neutrinos and anti-neutrinos coming from these distant sources—and there are such possibilities in the future—then we could say for sure whether the distant galaxy is made up of particles or anti-particles.

Although purely from the point of view of experimental evidence we cannot assert that distant galaxies are made up only of matter, astronomers generally agree that matter is dominant in the whole of the Universe. This raises a fundamental question: "Particle physics does not favour particles over anti-particles. Thus it is reasonable to expect that when the Universe was born there were equal numbers of both particles and anti-particles. What then happened subsequently to eliminate almost all the anti-particles leaving the particles as survivors? In other words, how did the Universe go over from a state of particle–anti-particle symmetry to one of asymmetry?" That is what we shall now be concerned with. (I am here reminded of time-reversal symmetry. As I explained in ***A Hot Story***, at the level of particle physics there is time-reversal symmetry, but in the macro world there isn't. We have something similar here. The Universe was born with matter–anti-matter symmetry, but later acquired an asymmetry.)

A long time ago when Dirac gave his hole theory that was the forerunner of the prediction of the existence of the positron, which, remember is the first anti-particle that we got to know, the ever-suspicious Pauli wrote in 1933 to Heisenberg, "I do not believe in the hole theory because I would like to have the asymmetry between positive and negative electricity in the laws of Nature. It does not satisfy me to shift the empirically established asymmetry to one of the initial state." What Pauli was saying was, in today's language, effectively just this: "You can't simply get away with it all by declaring that the Universe was created with an asymmetry, with a bias, there being more of matter than anti-matter." In fact, in later years when the subject of matter–anti-matter asymmetry in Nature came to be actively discussed, most physicists rejected scenarios that *demanded asymmetry as an initial condition.* Pauli, with his famous gut feeling, rejected this idea right away, long before anyone else did and well before the subject became a hot topic.

Before we go deeper into the subject, I want you to first appreciate that when we are talking of matter–anti-matter asymmetry, we are basically talking of baryons. After years of speculation, we now have some idea about why such an asymmetry exits. Of course, many details of the solution to the puzzle are missing right now, but it is fair to say that we do have an overall idea of why this partiality is to be found in Nature.

8.2 Origin of matter–anti-matter asymmetry

Our story begins in 1967 about three years after CP violation was experimentally discovered (see Volume 1). Andrei Sakharov of what was then USSR, wrote an important paper in which he essentially said that CP violation holds the clue to the asymmetry we are talking about. He said (effectively): "Listen, we don't have to push this asymmetry to an initial condition associated with the birth of the Universe. That is a pretty clumsy thing to do. Let us agree that when the Universe was born there was thermal equilibrium and therefore an equal number of particles and anti-particles. True, many of them would annihilate by collision but many more would also be born. In all this the baryon number would be conserved. (I am sure you would be able to figure that out!) Baryon number being conserved means that as long as there is thermal equilibrium, there would be equal numbers of particles and anti-particles. OK, into this comfortable state of affairs let us bring in CP violation. This happens when the Universe goes out of thermal equilibrium. Going out of equilibrium means essentially that the collision rate of particles in the Universe does not keep pace with the overall expansion rate of the Universe. What I mean is that thanks to the expansion, the particles have fewer opportunities for collision and the collision rate falls below a critical minimum required for maintaining thermal equilibrium. And that is when asymmetry sets in. So one does not have to imagine that the Universe was born with a bias."

Perhaps you are somewhat confused, so let me go over this a bit more slowly. Now what does thermal equilibrium really mean? It means that the particles of the system are colliding frequently and are therefore able to have a Maxwellian kind of velocity spectrum which is the index of equilibrium. To put it in the old kinetic theory language, the mean free path for collisions is small, and there are frequent encounters. The particles are thus 'closely in touch'. Now remember, the Universe is expanding and rather fast at that. Thus, a stage may come when particle collisions cannot take place fast enough. (What this means is that the particles and anti-particles stop colliding as frequently as they did earlier.) That is when it is no longer meaningful to talk of a temperature. Everyone talks freely of what the temperature of the Universe was when it was 10^{-12} s old, when it was 1 s old and all that, but have you heard anyone talk of today's temperature? There is something there, is there not? (But how come people talk of the temperature of the CMB or the microwave radiation detected by Penzias and Wilson? Think about it!)

Let us now go back to the early Universe, CP violation and all that. Equilibrium is lost due to expansion; CP violation then begins to operate, after which asymmetry also sets in. But would not CP violation require the violation of baryon conservation? Absolutely, and who says that baryon number conservation should not be violated? Is there any *proven* law which says that such violation cannot occur? In other words, loss of equilibrium, CP violation, and violation of baryon conservation would *together* do the trick!

This is essentially what Sakharov drew attention to. His idea was considered revolutionary at that time, especially because he invoked the violation of baryon conservation. Till then no one had dared to question this conservation law, but

Sakharov did. You must appreciate that Sakharov did not give a full theory—we simply did not know enough about particles and their behaviour at that time to attempt such a theory. Thus, what Sakharov gave was the *seed* of an idea, and an important seed it was. By the way, Sakharov later became a prominent public figure. This was not because of his scientific work but because of a political stance he took. You see, early in his career, he was one of the chosen few who worked on the hydrogen bomb project in USSR— that was because he was an expert in nuclear physics. But later, he became a political dissident and that was when he was thrown into prison. If it had been Stalin's days he would have been executed straightaway, but luckily for him, things had eased slightly by the time he started objecting to the policies of USSR. Thanks to the cold war then in progress, this dissidence immediately made Sakharov a hero in the West. After spending years in detention, Sakharov was finally released when USSR itself was about to collapse. He then went to live in America where he died later.

OK, Sakharov gave an idea; has anything been done after that? Definitely. Using the Standard Model many scenarios have been worked out for how the asymmetry actually came about. One thing that all such studies have revealed is that the SM by itself is not adequate for describing the asymmetry; modifications to the SM are definitely required. Many versions of the theory based on various types of modifications have been proposed; the final verdict is yet to be delivered. As they say in America, the jury is still out there, debating. Although the final word is yet to be said, we certainly have a rough scenario as to why there is this asymmetry today between matter and anti-matter in Universe, and that is what I shall now outline. It goes like this.

When the Universe was born, the temperature was astronomically high, more than 10^{28} K, and at around 10^{-35} s, the Universe was full of X particles and its anti-particle, namely, $\bar{X}$ (or anti-X). I hope you remember the X particle from Chapter 3 where I discussed proton decay. The X particle is quite heavy, and if it has to be created, very high energies are required. In turn, this means that the temperature of the Universe must have been very high; and at about 10^{-35} s the temperature was just about right which made it possible for X and anti-X to be created in large numbers.

Time passed and the Universe cooled below the threshold for the production of X. No more new X particles, and many of the already present X and anti-X were disappearing through decay into protons and anti-protons. See Fig. 8.1. But while this decay was taking place, something rather subtle was also happening, and that is the violation of baryon conservation. As we know from Chapter 3, baryon conservation is not an exact symmetry but an approximate one going hand-in-hand with CP violation, as was pointed out by Sakharov. Experts believe that CP violation which we now find to be so very rare, must have been common in the very early Universe, i.e., when the Universe was full of X and anti-X.

Granted that CP violation was in full swing then, and that baryon conservation was not a sacred rule. So what? Well, in such a situation *there is a slight asymmetry between the number of protons and the number of anti-protons produced.* The asymmetry may not be much but it is there nevertheless. Even if it is as small as 1 in 10^{15}, it can play a crucial role as we shall presently see.

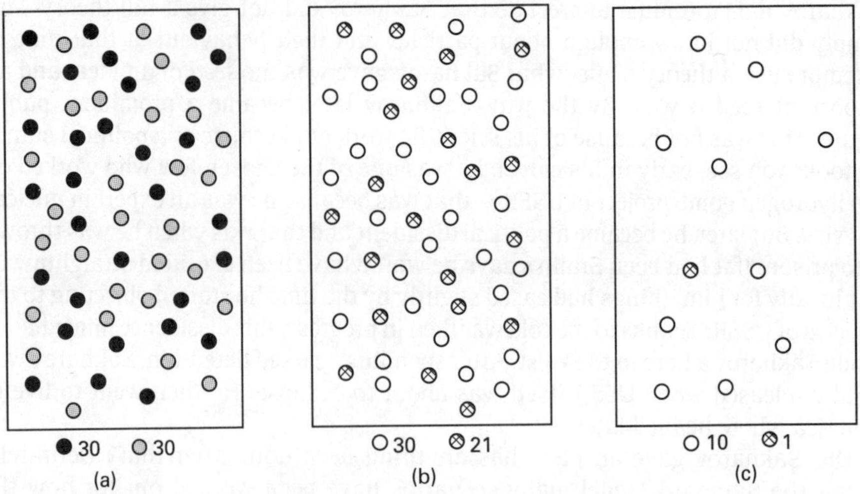

Fig. 8.1 *This figure illustrates schematically the emergence of asymmetry between matter and anti-matter in the cosmos. At very early times (~ 10^{-35} s), there were equal numbers of X and $\bar{X}$ particles. Subsequently, the X and $\bar{X}$ decayed, giving rise to protons and anti-protons. Thanks to CP violation, this transition was accompanied by a small asymmetry, meaning that at around 10^{-32} s or so, the protons were slightly in excess compared to anti-protons. Between 10^{-32} s and 10^{-4} s say, most of the protons and anti-protons disappeared through annihilation. As a result, after about 10^{-4} s, the number of surviving protons was far in excess of the number of surviving anti-protons. Illustrating all this, in (a) we have 30 X and 30 $\bar{X}$; in (b), we have 30 protons and 21 anti-protons while in (c), we have 10 protons and just one anti-proton.*

You may at this point ask, "Wait a minute. You say that at time 10^{-35} s the Universe was full of X and anti-X. Why did they not annihilate themselves?" Good question, and let us look at the whole scenario a bit more carefully. Remember, there can be annihilation only when the particle and the anti-particle actually collide. A merciful Providence did not give enough time in the beginning for all of the Xs and anti-Xs to collide and actually vanish. Sure, some must have, but not all. Meanwhile the temperature decreased, and thanks to their decay the Xs finally made their exit, leaving protons and anti-protons, with a slight excess of protons due to CP violation. Now these protons and anti-protons (and there were plenty of them) had plenty of time for collision than did the X particles, and sure enough a large number of them (protons and anti-protons) paired off and vanished due to annihilation. At the end of this mass annihilation there were only a relatively 'few' survivors—protons and anti-protons, that is. More protons survived because CP violation favoured a slight excess of particles over anti-particles. Mind you, the excess at the start was very tiny, *but that was enough to imprint an asymmetry permanently on the Universe.* From then to now, the Universe has been carrying that signature of asymmetry. All this is summarised in Fig. 8.1.

8.3 Chapter summary

Physicists believe:

- When the Universe was born, particles and anti-particles were created in equal numbers.
- At $t = 10^{-35}$ s, there was thus an equal number of X and anti-X.
- X and anti-X particles existed then because the temperature T was high enough for their creation ($T \sim 10^{28}$ K).
- The X and the anti-X collided and annihilated each other to some extent, but not entirely. Meanwhile, more X and anti-X were also being produced because the temperature was high.
- Thereafter T decreased and the production of X and anti-X also stopped.
- At this stage, the population of X and anti-X began to decrease not on account of collision but through decay into protons and anti-protons. Along with it, CP symmetry was also violated. This favoured more decay for X rather than anti-X. Net result: more protons were formed than anti-protons.
- Following this there was plenty of time for the protons and anti-protons to collide. This collision wiped out almost all the anti-protons; only the protons that were in excess survived. That excess, which is a measure of the asymmetry, is a very small number of the order of 1 in 10^{10} or even 1 in 10^{15} as some experts think.
- And that matter–anti-matter asymmetry created a long, long time ago and has been around ever since.

This is a rather hand-waving overview of the story of why the Universe today is filled almost entirely with matter rather than anti-matter. Particle physics and early Cosmology together provide a framework for understanding how this asymmetry came about from a situation wherein perfect symmetry prevailed (just after birth).

The description I have given here is typical of the state of many areas of Cosmology. We have only rough pictures. If you see the original papers, true, they would be full of respectable looking equations and all that; but that does not make the picture any less fuzzy! You see, we still don't quite have a GUT, we still have not quite united gravity with the other forces, and so on. Thanks to such uncertainties, there is a proliferation of ideas, and some fuzziness is unavoidable as one starts talking of very short times and very small distances.

One thing everyone is sure of: CP violation which was discovered in the case of kaons many decades ago holds the important key. So experts are looking at various aspects of this CP violation, both theoretically and experimentally. Such investigations, it is hoped, would offer precise clues as to how exactly one must modify the SM. With every passing day, somewhere or the other, some question or the other is getting resolved, leading to greater clarity at least in that limited area. One day, all this may suddenly lead to an unexpected breakthrough, and after that the fog would suddenly lift like the mist in the morning Sun. Such a thing happened when QM was discovered, this happened again with QED, and also subsequently with electro-weak unification. We just have to be patient while the pundits are struggling! Climbing the Everest is not easy!

9

THE HUNT FOR PRIMORDIAL DEUTERIUM

Soon after the Big Bang the light elements were formed. Good estimates can be made of the relative abundances of these primordial elements, provided one has a good estimate of the overall matter density. Physicists have reversed this game: measure the relative abundances and then use that for estimating the density. Why bother about the density? Because its value will decide the future fate of the Universe. Deuterium is a particularly good light element to focus on because its relative abundance is very sensitive to the value of density. This chapter gives a brief account of the study of abundance of primordial deuterium.

The Hunt for Primordial Deuterium

The cosmos is full of wonderful things for astronomers to observe. At roughly one second after the Big Bang, weak interaction froze out; quarks had already combined to form protons and neutrons. All the matter one sees in the Universe has come from this initial resource bank. Subsequently, the very first neutral atoms of the light elements formed, and this took something like anywhere between 100,000 to 300,000 years or so. At that time the Universe was about a thousand times smaller than what it is now. In due course, these neutral atoms began to coalesce into gas clouds, which later evolved into stars. One thing you must note here is that while stars have recycled original matter into newer elements that did not form in the beginning, they did *not* create matter; all the matter that is in the Universe now was already present in the very early state, though mostly in the form of simple elements.

As you know, astronomers are very fond of studying stars and galaxies. But they are also becoming increasingly interested in tracking the remnants of the early Universe. Unfortunately, there are not too many of these since most of what was produced in the beginning underwent transformation later. However, there are a few things left still. The CMB is one of them and that is why it is studied avidly all the time. The other tell-tale signature consists of the primordial elements, hydrogen, helium, deuterium, and lithium. The latter hunt is particularly interesting because the Big Bang model provides estimates of how much of these elements were formed in the very beginning. There is one thing very interesting about these estimates. Let us say, in the beginning there were in all N atoms. We now ask: Of these N atoms, what was the abundance of ^{1}H, ^{2}H, ^{3}He, ^{4}He, ^{6}Li and ^{7}Li? These are among the very first elements to be formed. It turns out that these relative abundances depend very much upon what the matter density of the Universe was at the time these primordial elements were formed. For different values of that original matter density, one obtains different values of the relative abundances of these basic elements. Let us say we have a graph of these relative abundances versus density at the time they were originally formed. These estimates can be made on the basis of the Big Bang model. We know the expansion rate of the Universe. Therefore we have an idea of the present volume of the Universe. Hence, that graph of the relative abundance versus density pertaining to the early Universe can be converted into another graph that shows the relative abundances versus densities of the type that we can expect today. Such a graph is shown in Fig. 9.1. The data shown here are calculated using a model that makes the following basic assumptions:

1) The Universe is homogeneous and isotropic.
2) Einstein's Theory of General Relativity holds.
3) The CMB was generated by the Big Bang.

Together with a number of satellite assumptions, the curves of relative abundance versus current density have been determined for the light elements.

Of what use is such a graph? Well, let us say we are able to somehow measure the density of, say primordial hydrogen present in the Universe. Using the graph in Fig. 9.1, we can then guess the density of the Universe today. And that number is very important in the context of deciding what the future of the Universe would be. Note one very interesting feature about the graph in Fig. 9.1 The graph for deuterium varies very sharply with density. What this means is that small changes in density are reflected as big changes in the relative abundance of primordial deuterium. Thus the relative abundance of deuterium can offer a very sensitive measurement of the density, even more than other light elements. In other words, a study of the relative abundance of deuterium can, in principle, provide an accurate estimate of the density of the Universe today and thereby also give a good clue about the future of the Universe. That is the reason for the special hunt for deuterium on which we now focus.

Let me state all this once again in a slightly different fashion. One looks into space, especially in the vast region between stars, and checks on the deuterium density there. If any deuterium is present, it most likely is primordial in origin. While looking out for deuterium, one also looks in the same region for the hydrogen density. This way, one gets the deuterium-to-hydrogen ratio. One can suppose that the hydrogen in that region is not having any remnants from stars and is primordial in nature. From all this, one can get an idea of the overall matter density, which is important in the cosmological context.

This raises the question: How does one go about detecting primordial deuterium? This is explained in Fig. 9.2. Consider first ordinary hydrogen. We see from Fig. 9.2 that when the spin of the electron in the hydrogen atom flips, an electromagnetic radiation is emitted at a wavelength of 21 cm. This 21 cm radiation from hydrogen is very famous in radio astronomy. It has been extensively studied and valuable information about the presence of hydrogen in

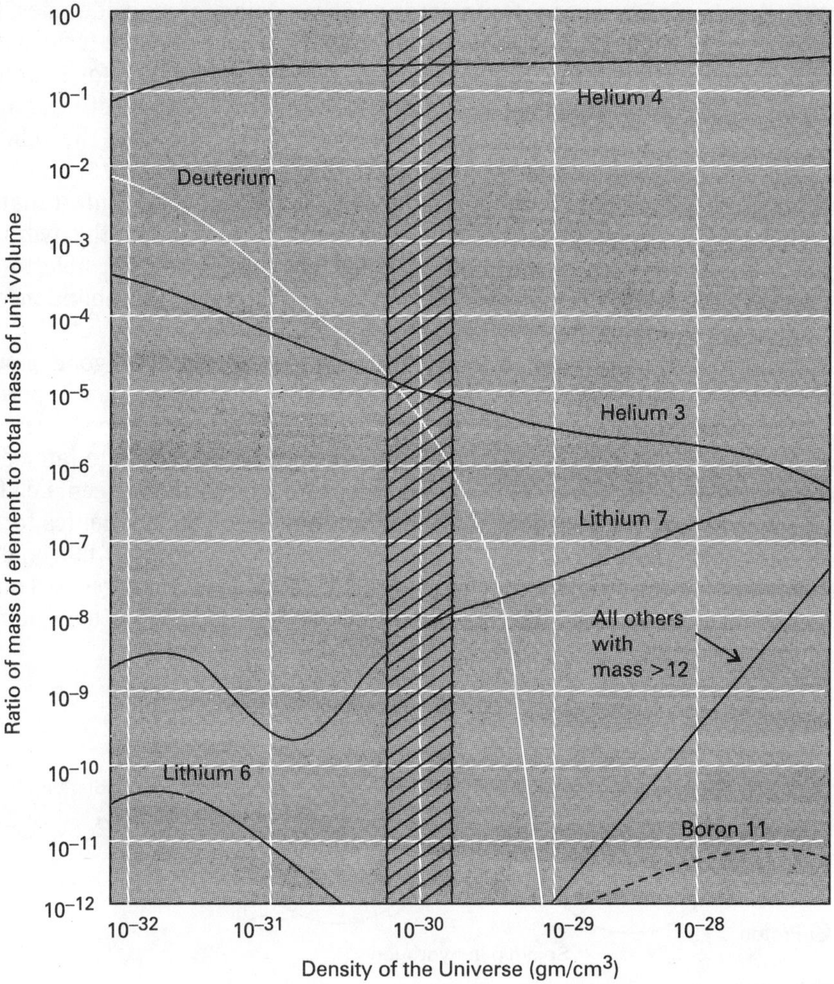

Fig. 9.1 *This figure shows the variation of the relative abundance of the light elements as a function of the current matter density in the Universe. This graph is generated as follows. Use the Big Bang model. Add other considerations. Calculate relative abundances versus density for the early Universe. Now bring in expansion of the Universe and recalculate these abundances for the current size (for each density of the early Universe, there would be a corresponding density at the present time). This is the way the above graph is generated. How does one use it? Measure relative abundances today. Use the graph to check which value of density is consistent with the observed relative abundances. That is the value of matter density one is after. The vertical band shows the narrow range of densities that is consistent with the observed relative abundances.*

space has been obtained. For deuterium, the corresponding wavelength is 92 cm. So you would think that all one has to do is to track the 92 cm radiation and thus obtain the information that one wants.

Well, that is exactly what astronomers have been doing for many decades. But the problem is that detection of the deuterium signal has not been easy. Incidentally, when astronomers tune into deuterium they also try to get data for hydrogen from the same source; this way they can get a value for the ratio of primordial deuterium to primordial hydrogen. If one has a good value of the relative abundances of primordial hydrogen and deuterium, then using Fig. 9.1, one can obtain a pretty good value of the matter density in the Universe. Unfortunately, one problem that astronomers face in all such investigations is that the hydrogen signal is so strong that it saturates. In other words, they are not able to say precisely what the deuterium-to-hydrogen ratio is because the abundance of deuterium is very much smaller than that of hydrogen.

So what's the finding? Well, it looks like one does not yet have a good enough value for the deuterium-to-hydrogen ratio. Historically, this was an important number to measure in order to get an estimate of the density of the Universe. A good estimate of the density was required to determine the ultimate fate of the Universe: whether the Universe is closed or open, or exactly in between, i.e., flat. Since then, despite the poor precision of deuterium data, good estimates for the other light elements such as ^{3}He, ^{7}Li, and of course hydrogen itself has enabled astronomers to narrow down the value of the density of the Universe to a small range of uncertainty (see Fig. 9.1). Anyway, this density is what astronomers were really after.

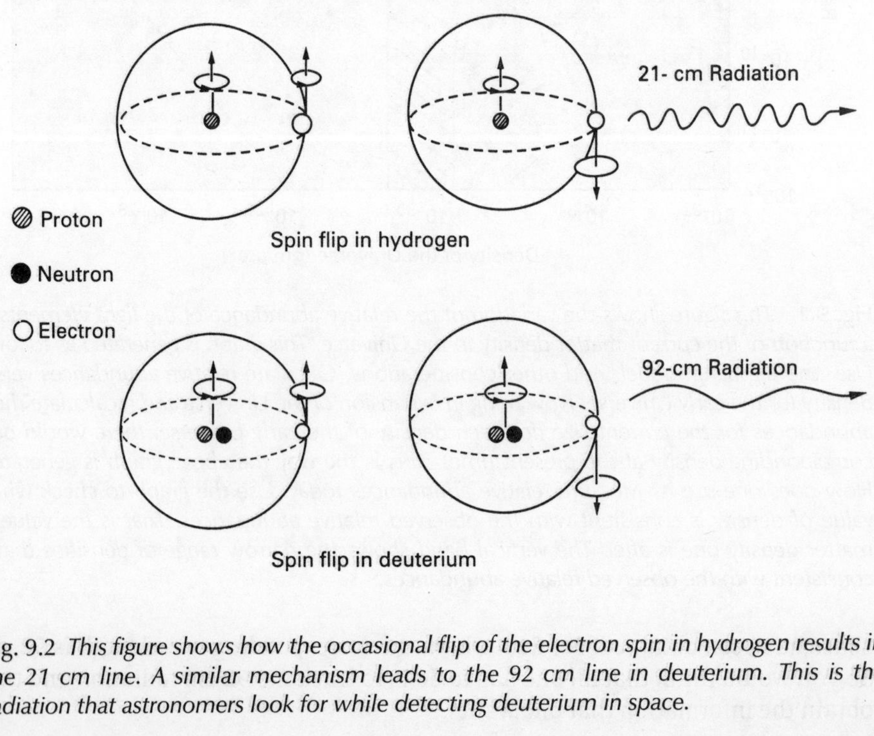

Fig. 9.2 *This figure shows how the occasional flip of the electron spin in hydrogen results in the 21 cm line. A similar mechanism leads to the 92 cm line in deuterium. This is the radiation that astronomers look for while detecting deuterium in space.*

At the and of it all, what is that the astronomers are saying? Too little matter for the Universe to close! Many other experiments have also clearly indicted that there is too little matter in the Universe to make it closed or even flat; that is, the value of Ω based on matter density studies is very much less than 1. On the other hand, other experiments have indicated that the value of Ω is close to 1, which means that the Universe *is* flat. So we have a big puzzle: How can the Universe have low density of matter and still be flat? That topic is reserved for a later chapter. As far as the deuterium issue is concerned, I am not too sure if it has any longer the importance it once had because other experiments have intervened and clarified the situation regarding the value of Ω, which really is the key issue.

10

ARE THERE MONOPOLES?

This brief chapter deals with the question of magnetic monopoles, that is, the magnetic counterpart of electric charge. They don't appear to exist in Nature, but is that really true? Could it be that they exist but we have missed them thus far? That question raised by particle physics is briefly discussed in this chapter.

This chapter is about magnetic monopoles. To understand what a magnetic monopole is, we start with the familiar electric charge. Let us say we have two electric charges separated by a short distance, one of them being positive and the other negative. We then have what is called an electric dipole, composed of two electric monopoles. If we have two electric dipoles, we get an electric quadrupole; two quadrupoles make an octupole, and so on.

The magnetic dipole can be understood by analogy with the electric dipole. To understand the magnetic dipole, let us consider a bar magnet as in Fig. 10.1 (a). As we all know, the magnet has two poles, a north pole and a south pole. The north and the south poles together make up a magnetic dipole. Suppose the bar magnet is cut into two pieces as in Fig. 10.1(b). We then get two smaller bar magnets each having a north and a south pole. No matter how much we try, we *cannot* ever isolate a magnetic north pole or a magnetic south pole. Nature seems to have bound the opposite magnetic monopoles inseparably, and no one can tear them apart. In contrast, electric charges can stand alone; that is to say, an electric dipole can be physically separated into two electric monopoles, meaning we can have an isolated positive charge and an isolated negative charge. To sum it all up, while electric monopoles (meaning charges) can stand alone, magnetic monopoles or magnetic charges apparently cannot. By the way, in what follows, the term monopole would always mean the magnetic monopole, though from the language point of view, the term monopole could equally well be applied to an isolated electric charge (which is nothing but an electric monopole).

Getting back to physics, there somehow appears to be a kind of bias or asymmetry in Nature as far as electric and magnetic charges are concerned. Maxwell who gave us the equations subsequently named after him recognised this, and thus the Maxwell's equations that we study *have an in-built asymmetry.* However, scholars have for long noted the following. To appreciate what they said, let us first write the equations as they apply to in vacuum. We then get,

$$\text{Curl}\ \boldsymbol{B} = \partial \boldsymbol{E}/\partial t$$

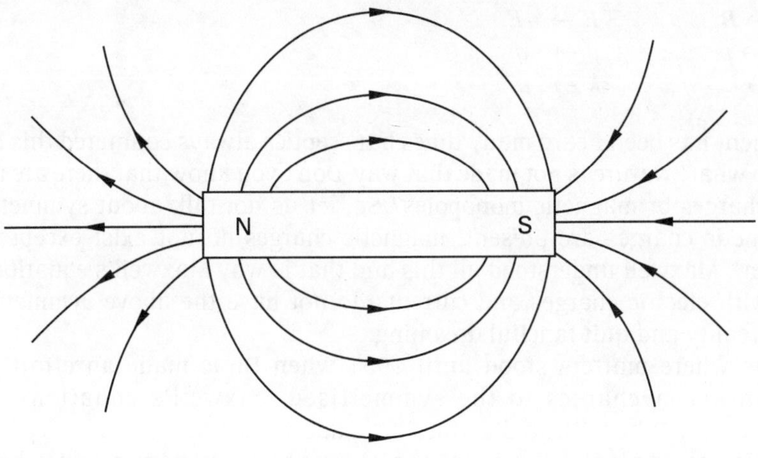

(a)

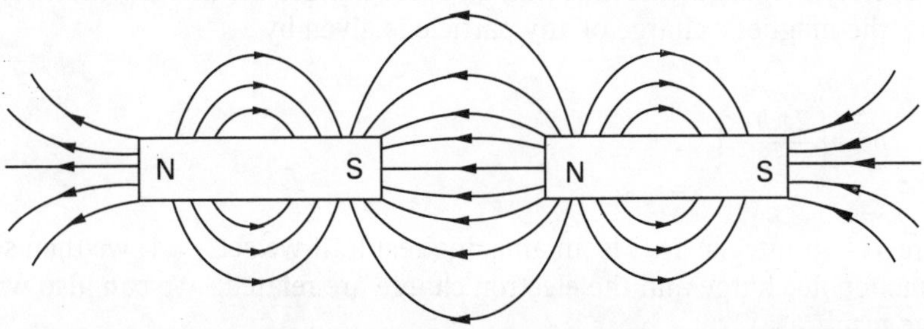

(b)

Fig. 10.1 *Here (a) shows the lies of force due to a bar magnet. As per our usual convention, the magnet has a north pole and also a south pole. Suppose the magnet is cut into two to make two pieces. The two pieces then each behave like a bar magnet, as shown in (b). The message of Nature is that unlike electric dipoles which can be split into electric monopoles, the magnetic dipole cannot be split in a similar manner.*

$$\text{Curl } \boldsymbol{E} = -\,\partial \boldsymbol{B}/\partial t.$$

These equations are symmetric in the sense that if we make the interchanges $\boldsymbol{E} \rightarrow \boldsymbol{B}$, and $\boldsymbol{B} \rightarrow -\boldsymbol{E}$, we get the same pair of equations that we started with. OK, Maxwell's, equations are symmetric in vacuum, so what? Scholars then said, "Supposing magnetic charges or magnetic monopoles exist just as electric charges do. We then have magnetic currents in addition to electric currents. Let us denote the electric charge by q and the magnetic charge by p. Further, let $\boldsymbol{j}$ and $\boldsymbol{k}$ denote respectively the electric current and the magnetic current (that is, the current due to the motion of the magnetic charges). The (extended Maxell's) equations resulting from the addition of these two types of charges are symmetric once more". More explicitly, what the scholars said was that the modified equations become *symmetric* under the interchanges

$$\begin{array}{ll} \boldsymbol{E} \to \boldsymbol{B}, & \boldsymbol{B} \to -\boldsymbol{E}, \\ q \to p, & p \to -q, \\ j \to \boldsymbol{k}, & \boldsymbol{k} \to -j. \end{array}$$

This argument has been heard many times but sceptics always countered this by saying, "So what? Nature is not made that way. Don't you know that there are no magnetic charges or magnetic monopoles? So, let us not talk about symmetry when magnetic charges are present; magnetic charges do not exist except in imagination". Maxwell understood all this and that is why Maxwell's equations complete with electric charges and currents do not have the above symmetry. Stick with reality and quit fanciful dreaming.

This was where matters stood until 1931 when Dirac made an effort to apply quantum mechanics to the symmetrised Maxwell's equations as discussed above. When Dirac did this, he made a surprising discovery. He found that if one admits that there is even one magnetic monopole somewhere in the Universe, then electric charge is quantised! More explicitly, if e is the basic unit of electric charge (meaning that all other charges are integral multiples of e), then p the magnetic charge of any particle is given by

$$p = n\left(\frac{2\pi\hbar c}{2e}\right),$$

where n is an integer. This is an amazing result. If we set $n = 1$, we then see that the monopole charge and the electron charge are related. We can also write the above result as

$$q = n'\left(\frac{2\pi\hbar c}{2p}\right), \qquad n' = \text{an integer.}$$

This tells us that electric charge gets quantised if monopoles exist in the Universe. This would also explain, for example, why the proton has the same magnitude of electric charge as the electron. Note that application of QM does not straightaway lead to monopoles and all that. On the other hand, if one 'puts the monopole in by hand' (that is to say, postulates their existence), then charge quantisation immediately follows. Since charge is in fact found to be quantised in Nature, perhaps there are monopoles in Nature after all—that was Dirac's basic argument. The message was: "Let us not dismiss magnetic monopoles in haste just because we have not come across them".

A word or two now about the peculiar nature of Dirac's monopole. It all came out of an analysis of the usual vector potential $\boldsymbol{A}$ of electromagnetic field theory, with particular reference to the monopole. Dirac found that if one has a monopole, which, say is at the origin, then the associated vector field $\boldsymbol{A}$ is defined everywhere except along a line stretching all the way from the origin to

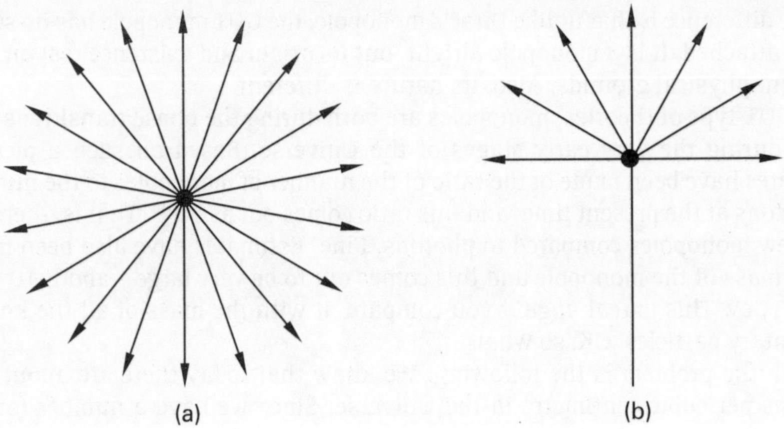

Fig. 10.2 *This figure illustrates the nature of the Dirac string. In (a) are shown the field lines that we would normally expect for a (positive) magnetic monopole located at the origin. But due to various complications it does not work out that way. Rather, the situation is as in (b), as Dirac discovered for the first time. Here we see a string extending from the origin all the way to minus infinity. This string represents a line singularity, and in this region the field is not defined.*

minus infinity! It is as if a long string is attached to the monopole, and in fact this string is sometimes called the ***Dirac string*** (see Fig. 10.2). Whereas we would normally expect the field lines of the magnetic monopole to be as in part (a) of that figure, it is actually as in (b) which shows that in the region within the string the field is undefined.

Following Dirac's prediction, several attempts were made to detect the magnetic monopole. For a while a claim was even made that the monopole had been detected, but this claim did not withstand scrutiny and no one has ever succeeded in actually detecting a monopole thus far despite the many methods that have been used. Dirac made an estimate of the mass of the monopole and concluded that it was expected to be very heavy. Monopoles could therefore not be easily produced as pairs, like one produces electron–positron pairs, for example. What about monopoles occurring naturally? May be they are around somewhere and that is what has motivated people to look for them. Dirac's monopole was not taken seriously by large sections of the theoretical physics community (except perhaps as an 'interesting' result) until 1974. That year, t'Hooft in Holland (you met him in Volume1) and Polyakov in Moscow found that in certain types of GUT models, monopoles find a place naturally—in fact, their presence is necessary. That is to say, one does not have to postulate their existence; rather, they follow from basic gauge principles that are a feature of GUT. In other words, t'Hooft and Polyakov found that monopoles exist quite naturally in certain types of gauge theories and do not have to be 'put in by hand', as in Dirac's analysis. Naturally, with this discovery the monopole acquired much respectability in particle-physics circles.

One difference is that unlike Dirac's monopole, the GUT monopole has no string or tail attached. It is a monopole alright, but its origin and existence rest on very different physical grounds. Also, its nature is different.

In GUT-type of theories, monopoles are born during the phase transitions that occur during the very early stages of the Universe. Based on such a picture, estimates have been made of the ratio of the number of monopoles to the number of photons at the present time, and this ratio comes out as 10^{-10}. That is, there are very few monopoles compared to photons. Fine. Estimates have also been made of the mass of the monopole and this comes out to be very large—about 10^{-8} gm or 10^{16} GeV. This is real *huge* if you compare it with the mass of all the known elementary particles. OK, so what?

Well, the problem is the following. We know that today there are about 400 photons per cubic centimetre in the Universe. Since we have a number for the ratio of the monopole density to photon density, we can calculate the number of monopoles per cubic centimetre. From this we can estimate the mass contributed by all the monopoles in the Universe, since we have a good idea of the volume of the Universe at the present time. When this calculation is done, it is found that the mass of the Universe comes out to be astronomically large, I mean it is something like 10^{13} times larger that what we know the mass of the Universe to be, not withstanding all the experimental uncertainties. This is a phenomenal discrepancy, and clearly something is wrong somewhere.

Pundits have been worrying about this for quite a while and have been desperately searching for a mechanism that would make most of the monopoles 'just go away'. Luckily, one such model has been found, the *Inflationary Model* of the Universe. That model would be discussed in a later chapter. What experts now say is: "All the original monopoles produced have been inflated away, and the monopoles we are now left with are those that were produced just after inflation or towards the very end of the inflation era." The number of such monopoles in existence is expected to be very much smaller than the original estimate we had. I am sure this inflation business must be puzzling you a lot; but don't worry, for I shall be telling you all about inflation in a later chapter. Meanwhile, it is enough to remember the following: Monopoles are quite possibly there, though very scarce. They are there on account of GUT, having been formed during the early phase of the Universe. True, we do not as yet have an official GUT model, but even so I think we can safely bet on monopoles!

11

ELUSIVE BUT IMPORTANT AND FASCINATING

In this somewhat long chapter is given a comprehensive overview of the current status of neutrino physics, both in relation to the particle physics aspect and the astrophysical aspect.

11.1 Introduction

In Volume 1 we took a good look at the neutrino. Interestingly, this tiny and rather elusive particle is playing today a key role not only in particle physics but in Cosmology as well. In this chapter I shall be offering you some more glimpses of the neutrino story. Part of it would be pure neutrino physics and part of it would be related to the cosmology aspects. Of course, a clear separation like this is really not possible; however, for operational reasons we shall to try look at these matters separately, wherever convenient. Here is a glimpse of what we shall be considering.

To start with, we note that for a long time there has been an unanswered question: Is the neutrino a Majorana particle or a Dirac particle? That is to say, is the neutrino its own anti-particle (which is what Majorana said) or is it otherwise (which is what Dirac said)? Experiments that would answer this question are in progress, and that is what we shall discuss first.

Next, there is the question of neutrino mass. In Volume 1 we sort of took this mass to be zero; but is it really so? Models (such as the Standard Model, for example) declare the mass of the neutrino to be zero but what is the reality? Actually, from a physics point of view, there is no fundamental reason why the mass should be exactly zero. In fact there are models going beyond the SM that demand the mass to be finite. Once again the question arises: What does Nature say? We now more or less have the answer and this too would come up for discussion, particularly in relation to a question closely related to non-zero mass, namely the inter-conversion of neutrino flavours. As you know from Volume 1, we have three neutrino species, the electron neutrino, the muon neutrino and the tau neutrino. Current theoretical models suggest that over a period of time one species can convert to another. This phenomenon called *neutrino oscillation* can occur only if the neutrino mass is non-zero. Naturally, a big effort has been mounted to detect neutrino oscillations and this would receive quite a bit of attention in this chapter.

In Volume 1, I described the heroic attempt made to actually detect the neutrino. Since then, things have advanced considerably. Currently, not only are we able to

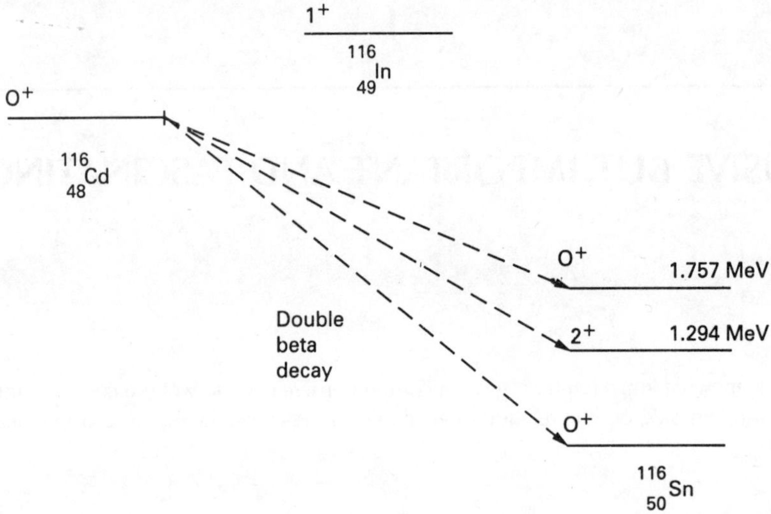

Fig. 11.1 *Shown here are the low-lying nuclear levels of three nuclei that form an isobaric triplet (same mass number – in this case 116). These are the levels crucial for double beta decay. The spins and parities of the concerned levels are also shown. Decay from cadmium to indium is energetically forbidden but decay from cadmium to tin is not. However, this calls for double-beta decay. We would be concerned mostly with decay to the ground state of tin, but decays to the two higher states of* ^{116}*Sn are also possible.*

detect neutrinos coming from the Sun, but indeed even those coming from supernova explosions! All this has opened up the exciting possibility of neutrino astronomy; the chapter would therefore be rounded off with some references to this new vista.

11.2 Double beta decay

Double beta decay is a rare decay process in which a nucleus decays by emitting two electrons simultaneously. The manner in which this can occur can be understood by referring to Fig. 11.1. Here we see that the isotope $^{116}Cd_{48}$ cannot decay by beta emission to its neighbour $^{116}In_{49}$ since such a decay is energetically forbidden. However, the cadmium isotope could decay to a nucleus two steps away, namely $^{116}Sn_{50}$, but this calls for the emission of *two* electrons. Thus, if the cadmium isotope decays to the tin isotope, we say there is double beta decay.

Double beta decay is a second-order process mediated by weak interactions, and the half life is consequently very long. Now it turns out that double beta decay can occur in various ways, depending on whether lepton number conservation is respected or violated. To understand all this, we start with what we already know, namely the way the neutron decays. We have,

$$n \rightarrow p + e^- + \bar{\nu}.$$

Keeping this in view, let us consider a transition where a nucleus (*A*, *Z*) changes into a nucleus (*A*, *Z*+2) as in Fig. 11.1. We can say that this happens due to two neutrons transforming as above into two protons and thus write:

$$n_1 \rightarrow p_1 + e_1^- + \bar{\nu}_1 ,$$

$$n_2 \rightarrow p_2 + e_2^- + \bar{\nu}_2 .$$

In effect, this gives us

$$(A, Z) \rightarrow (A, Z+2) + e_1^- + e_2^- + \bar{\nu}_1 + \bar{\nu}_2 . \quad (1)$$

Notice that in the above, double beta decay occurs with the emission of two anti-neutrinos. Now the transformation of two neutrons into two protons can also occur in another way as below. We have first

$$n_1 \rightarrow p_1 + e_1^- + \bar{\nu}_1 ,$$

which is just as before. Let us suppose the neutrino emitted in this reaction interacts with a neutron. This gives

$$n_2 + \nu_1 \rightarrow p_2 + e_2^- .$$

You might wonder: "This reaction is for a neutron interacting with a neutrino whereas in the first reaction we had an anti-neutrino being produced. Where did this neutrino suddenly pop up form? Something funny going on here!" Not really, provided the neutrino is its own anti-particle. In other words, the transition

$$(A, Z) \rightarrow (A, Z+2) + e_1^- + e_2^- \quad (2)$$

is possible if the neutrino is of the Majorana type. Notice that in this case there is double beta decay *without* the emission of any neutrinos. Hence this process is called *neutrinoless double beta decay* ($0\nu\beta\beta$). (Pundits may frown over this description since I have glossed over the complications caused by parity non-conservation. I submit that for our present purpose, this is OK.)

Take a minute off and compare equations (1) and (2). You will find, remembering all that was said about lepton numbers in Volume 1, that whereas (1) respects lepton number conservation, (2) violates it since the lepton number changes by 2 in the transition. What it means is that if experimentally one is able to observe neutrinoless double beta decay, it kills two birds with one stone. Firstly, it would establish that the neutrino is a Majorana particle. Secondly, it would establish that lepton number is not conserved, which is contrary to what SM demands. But models going beyond SM are ready to violate lepton conservation, and so the detection of neutrinoless double beta decay would be one up for such models.

OK, if it is all that important, then why don't people go ahead and detect neutrinoless double beta decay? Well, people have been trying, and in fact trying very hard. But detection is not easy since the half-life for the decay is estimated to be anywhere from 10^{22} to 10^{25} years. For comparison, remember that the current age of the Universe is only about 10^{10} years!

> ***Question:*** How does one distinguish double beta decay from the emission of neutrinos from neutrinoless double beta decay?

There is in fact a direct way of doing this, as would be clear by referring to Fig. 11.2. To understand this figure, we must first look at equations (1) and (2). In the former there are four emitted particles, two electrons and two neutrinos, while in the latter there are only two electrons. You will agree that the energy released in the decay must be shared between all the emitted particles. In the case of neutrinoless double beta decay (0νββ), the two electrons would carry away, between themselves of course, all the available energy. Hence, if one could measure the *sum* of the energy of the two emitted electrons, one would get a sharp line as in Fig 11.2. On the other hand, in the case of equation (1) the two emitted electrons would have to share some of the decay energy with the two neutrinos. In this case, this division or partition of energy if you like, can take place in an infinite variety of ways. The net result is that the electron sum energy spectrum is continuous as shown in Fig. 11.2. I am sure you would remember a similar discussion about the energy spectrum of the emitted electron that took place when the neutrino was first proposed (see Volume 1, Chapter 5).

OK, where do we stand today as far as experiments are concerned? Since the stakes are high, there has naturally been a lot of effort to detect (0νββ) decay. It is a very tough experiment because the half life is so large. You may say: "What is the big deal about the long half life? The time scales involved in proton decay are a million if not a billion times longer." True, but you see in the case of the proton decay we can make the sample quite large. We are after all dealing with water; no doubt very pure water and all that, but still it is water. In this case we need a particular isotope, and to get that in thousands of tons is not possible. That is what makes this experiment almost as challenging as the detection of proton decay.

Let us now see how people try to detect this rare process. There are two ways of detecting (0νββ) process—the so-called passive method and the active method. In the passive method one prepares the source in the form of a foil, say, for example we could think of a foil of ^{118}Cd. This foil is then placed between two very sensitive and high-resolution electron detectors. And one then waits! There would be millions and millions of background events and in the midst of this haystack one has to detect the needle that corresponds to the sharp peak of Fig 11.2. What about the second type of experiment? Here, one embeds the source into the detector. For example? Well, the germanium experiment is a good illustration. ^{76}Ge is one of the candidates for neutrinoless double beta decay. Using expensive and time-consuming isotope separation methods, a suitable quantity of ^{76}Ge is first produced. Next, this is mixed with normal germanium and a germanium detector is produced. This detector is heavily shielded, and thereafter one patiently waits and watches.

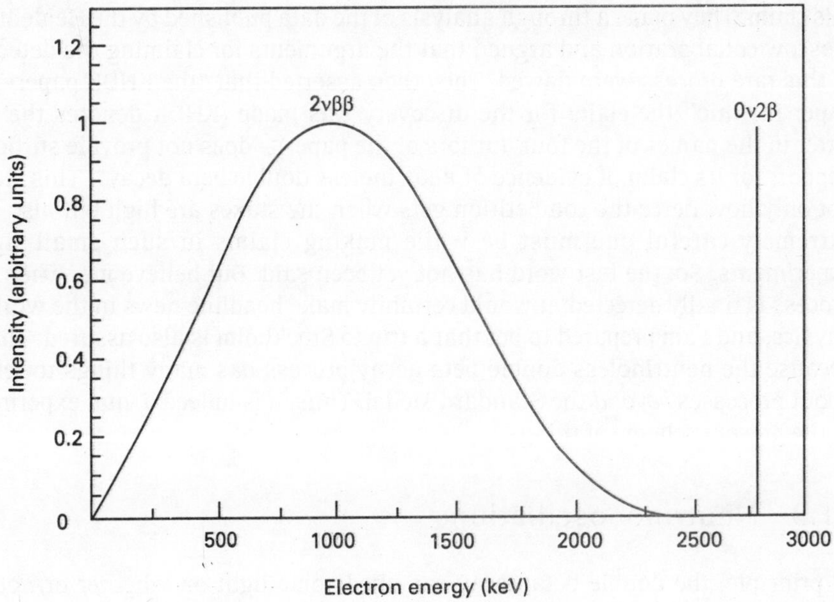

Fig 11.2 *Double-beta decay can occur in two substantially different ways, with the emission of two neutrinos or without the emission of any neutrinos. From neutrino-physics point of view, neutrinoless double beta decay is far more interesting. This figure shows how the two decay processes can be distinguished. Suppose one is able to catch both the electrons emitted in the decay and measure the sum of their energies. The electron-sum energy spectrum for the decay of* ^{116}*Cd would look as sketched above. Clearly, the signatures for decay with and without neutrino emission are so strikingly different that the two processes can in principle be distinguished.*

Just to give you a feel for what is involved in this type of experiments, let me mention a few details about two experiments based on ^{76}Ge now in progress. They are (1) the Heidelberg–Moscow experiment and (2) the IGEX collaboration. In the Heidelberg–Moscow experiment, there are five high-purity germanium detectors. The germanium used is enriched in ^{76}Ge to about 86%. In practical terms, what it means is that of the total mass of 11.5 kg of germanium in the detector, the active mass of ^{76}Ge is about 11 kg! The detector is kept in an underground laboratory, and all sorts of precautions are taken to shield against background events. The background rate has been brought down to about 0.06 counts per year per kg per keV—that should give you an idea of how careful these experimenters are. The other experiment is running in Spain. This too is a big collaboration and is running under the ground, below a mountain. In this experiment there are three germanium detectors, with the germanium enriched to 88%. There are a number of other such experiments, but I shall not go into them now.

So what is the score? Has double beta decay been positively detected? Not as yet. There was a flurry some years ago when the Heidelberg–Moscow group made such a claim, but very promptly the rival collaboration IGEX disputed this claim. They made a through analysis of the data published by the Heidelberg–Moscow collaboration and argued that the arguments for claiming the detection of this rare process were flawed. They then asserted that "the KHDK paper – the paper in which the claim for the discovery was made (KHDK denotes the first letter in the names of the four authors of the paper) – does not provide sufficient support for its claim of evidence of neutrinoless double beta decay". This shows not only how fierce the competition gets when the stakes are high but also how extremely careful one must be while making claims in such small-signal experiments. So, the last word has not yet been said. But believe me, when this process is finally detected, it would certainly make headline news in the world of physics, and I am prepared to bet that a trip to Stockholm is also assured. That is because the neutrinoless double beta decay process has many things to tell us about processes *beyond* the Standard Model. Thus, it is indeed a vital experiment in the present scheme of things.

11.3 Neutrino oscillations

In principle, the double beta decay can shed some light on whether or not the neutrino has a finite mass, but it seems that a better idea of that can be had from neutrino oscillations, to which we now turn. The idea of neutrino oscillation follows readily from basic quantum mechanics. You will recall (from ***QR I***) that according to QM, the time evolution of a stationary state $|\psi_k>$ is given by

$$|\psi_k> = |\psi_k(0)> \exp(-iE_k t/\hbar). \tag{3}$$

where E_k is the energy corresponding to the state $|\psi_k>$. The exponential factor tells us how the phase of the stationary state changes with time.

As explained in ***QR I,*** the time evolution of a non-stationary state is more complicated. For such a state, we can write

$$|\psi> = \sum_k a_k |\psi_k>, \tag{4}$$

where a_k are constants. Using (3), we find

$$|\psi(t)> = \sum_k a_k |\psi_k> \exp(-iE_k t/\hbar).$$

With this background, let us now turn our attention to the electron and muon neutrinos. For simplicity, we consider only these two neutrino flavour states. The basic assumption made in trying to explain neutrino oscillation is that ν_e and ν_μ are not stationary states but linear combinations of stationary states that are called *mass eigenstates.* Let us denote these mass eigenstates as $|\nu_1>$ and $|\nu_2>$. We can then write

$$|\nu_e> = \cos\theta\,|\nu_1> + \sin\theta\,|\nu_2>,$$

$$|\nu_\mu> = -\sin\theta\,|\nu_1> + \cos\theta\,|\nu_2>.$$

Consider now the state vector of an electron neutrino produced at time $t = 0$. Let the states $|\nu_1>$ and $|\nu_2>$ correspond to energies E_1 and E_2 respectively. From the above we have,

$$|\nu_e(t)> = \cos\theta\,|\nu_1> \exp\left(\frac{-iE_1t}{\hbar}\right) + \sin\theta\,|\nu_2> \exp\left(\frac{-iE_2t}{\hbar}\right). \quad (5)$$

We now ask: What is the probability $P(\nu_e, 0; \nu_\mu, t)$, that is, the probability for an electron neutrino ν_e born at time $t = 0$ to become a muon neutrino ν_μ at time t? This is given by

$$P(\nu_e, 0; \nu_\mu, t) = 4\cos^2\theta\,\sin^2\theta\left(\frac{\Delta m^2 t}{4p}\right), \quad (6)$$

where p is the neutrino momentum and $\Delta m^2 = m_1^2 - m_2^2$ is the mass square difference. The important point to note is that oscillations are intimately connected to the fact that neutrino has non-zero mass. It is of course an entirely different matter as to why the neutrino has a mass in the first place (remember, in the SM we took it for granted that the neutrino has a zero mass). As I shall describe a little later, there is now clear experimental evidence that Δm^2 is not zero. This means that we *must* go beyond the original SM and ask why the neutrino has a non-zero mass. We also need to examine what happened to some of the conclusions we drew earlier, asserting that the neutrino had a zero mass (see Volume 1, Section 5.10 and 11).

At this point let me just draw attention to the implication of result (6) above. Say a bunch of electron neutrinos is produced at time $t = 0$ at a point and that these neutrinos fly away. Then as time passes, many of these neutrinos would change their flavour as a result of the neutrino oscillation mechanism. They would each become a ν_μ. At a still later time, t_μ say, each of the original ν_e's would have become transformed into ν_μ's. After this, the reverse conversion starts with the ν_μ's changing back to ν_e's. Thus after a time $2t_\mu$, all the neutrinos would be back to the electron flavour – see Fig. 11.3. All this pertains to neutrino flight in vacuum. If neutrinos travel through matter, the interaction with matter will further enhance the oscillation mechanism. This is somewhat like optical birefringence but we shall not go into that. After much effort, neutrino oscillations have finally been detected recently, but the story does not start with an attempt to actually detect the oscillations. In a manner of speaking, it all started with solar neutrinos.

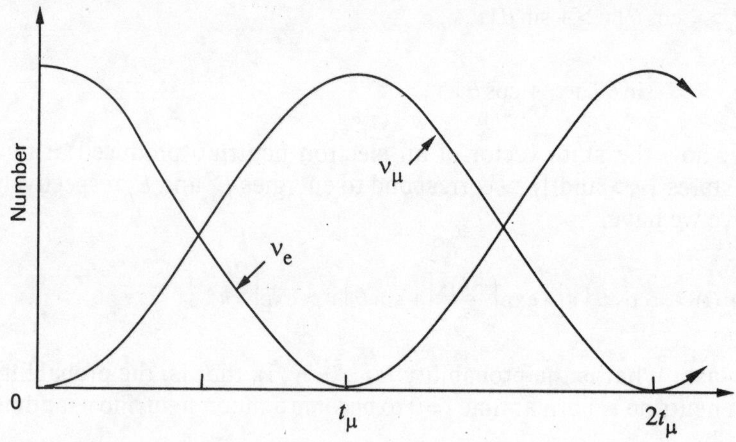

Fig. 11.3 *This figure illustrates what exactly neutrino oscillations mean. At $t = 0$, a bunch of electron neutrinos is produced and this beam flies away. During the journey, conversion would take place with some ν_e's getting converted into ν_μ's. This process would continue untill time t_μ when all the electron neutrinos are converted into the muon flavour. A reverse conversion now starts, and at time $2t_\mu$ all the neutrinos are back to their original electron flavour. This see-saw process goes on continuously.*

11.4 Neutrinos from the Sun

We all know that our Sun emits a lot of energy. What is the source of this energy? In the nineteenth century it was believed that the Sun was able to emit light and heat due to the release of gravitational energy. This energy production, people then felt, happened as the result of the gravitational contraction of matter within the Sun.

Around 1920, Sir Arthur Eddington (you have met him in ***Chandra***) realised that the energy of the Sun came not from gravity but from nuclear reactions. Eddington was in Cambridge, and you might remember from earlier books in this series that Cambridge was then the Mecca of nuclear physics, with Lord Rutherford as the presiding deity. So no wonder Eddington had the latest news about what was going on in nuclear physics.

OK, it is one thing to offer a conjecture but quite another to work out a *detailed* scenario of possible nuclear reactions taking place in the Sun, estimate the amount of energy released, and then compare this estimate with actual measurements available of the energy release. All this was possible thanks to some serious work by Hans Bethe (whose name also must be familiar to you). This work was done in 1938–1939. BOX 11.1 gives a summary of the principal reactions that lead to the fusion of protons into helium, accompanied by the release of neutrinos.

I want you to appreciate that Bethe's work was not only meticulous (as it always was in his case), but it also required daring imagination. Just reflect. All we can see is the surface of the Sun. We know heat and light come out from this surface, and we have an estimate, a good estimate in fact, of the surface temperature of the Sun—it is roughly 6000° C. We know the radius of the Sun and also its

BOX 11.1

Shown here are the principal nuclear reactions that lead to the fusion of protons into helium.

No.	Reaction	Neutrino energy (MeV)
1	$p + p \rightarrow {}^2H + e^+ + \nu_e$	0.0 – 0.4
2	$p + e^- + p \rightarrow {}^2H + \nu_e$	1.4
3	${}^2H + p \rightarrow {}^3He + \gamma$	
4	${}^3He + {}^3He \rightarrow {}^4He + 2p$	
	or	
5	${}^3He + {}^4He \rightarrow {}^7Be + \gamma$	
	then	
6	$e^- + {}^7Be \rightarrow {}^7Li + \nu_e$	0.38 – 0.86
7	${}^7Li + p \rightarrow {}^4He + {}^4He$	
	or	
8	$p + {}^7Be \rightarrow {}^8B + \gamma$	
9	${}^8B \rightarrow {}^8Be + e^+ + \nu_e$	0 – 15

mass. (Can you figure out how people determined these?) From this to nuclear reactions deep within the Sun occurring at a temperature of millions of degrees is one *huge* imaginative leap.

Bethe argued that the basic release of energy comes when protons combine or fuse and form helium atoms. Since Bethe was very strong and up to date in nuclear physics, he wrote down all the possible nuclear reactions that could lead to such fusion (for that is what it is), and calculated the energies. Now comes an important question. For such reactions to take place, the protons, etc., must have high energies. In the lab we can accelerate particles, make them acquire high energies and then produce nuclear reactions. Obviously, there is no accelerator inside the Sun. How then do the particles acquire such high energies for nuclear reactions to occur? The answer was easy to guess. Heat produced by the crushing force of gravity that acts in the centre of the Sun. The core temperature in the Sun is so high that simply on account of the heat, the particles start moving with *tremendous* velocity, and when particles collide with such velocities, nuclear reactions become possible. In short, Bethe showed that the Sun was a self-sustaining hydrogen bomb (where energy is released by the fusion of hydrogen into helium) —this long before anyone even thought of a hydrogen bomb!

I don't want to spend time on Bethe's wonderful achievement, important though it is. I am mentioning it for one very pertinent reason. It emerges from Bethe's analysis that if indeed the energy production in the Sun is as Bethe has described, then a lot of neutrinos must be produced. From BOX 11.2 we see that neutrinos, electron neutrinos

BOX 11.2

This is the story of Raymond Davis Jr., the saga of a man of tremendous perseverance and tenacity. Many decades ago, Davis decided that he would hunt for neutrinos coming from the Sun. He knew this would be a tough experiment but his point was: "People say that the Sun generates energy through thermonuclear reactions. What proof is there for this? If I can catch the neutrinos that would be generated in such reactions, then that would be proof indeed that the Sun is generating energy as Bethe proposed".

To detect the solar neutrino, Davis used a reaction suggested earlier by Pontocorvo and Alvarez. The reaction is

$$\nu + {}^{37}\mathrm{Cl} \rightarrow {}^{37}\mathrm{A} + e^{-}.$$

Essentially, thanks to this neutrino-induced reaction (called by the way as Pontocorvo–Davis reaction), an atom of radioactive argon with mass number 37 is produced. This argon isotope decays via electron capture with a half-life of 35 days. If you look at Fig. 11.4, you would realise that detection using chlorine misses the neutrinos emitted in the dominant (pp) reactions but picks up some of those from ^{7}Be decay.

What Davis did was to place a large tank at the bottom of a deep gold mine in South Dakota in the United States. Deep mines are a favourite for those who want to avoid cosmic ray backgrounds. This tank contained about 600 tons of a common cleaning fluid, tetrachloroethylene. Altogether there were about 2×10^{30} chlorine atoms in the tank. Based on an estimate of the number of neutrinos emitted by the Sun every second, Davis calculated that every month about 20 neutrino reactions would take place in his tank, i.e., about 20 argon atoms would be formed per month. Davis evolved a clever method of getting these twenty atoms out of the tank, and detecting them as they decayed. As someone said, this is more difficult than finding a particular grain of sand in the whole of the Sahara desert! Davis did this year after year and by 1994, he had collected about 2000 argon atoms! Imagine that! Just about two thousand atoms in roughly a quarter of a century! If that is not patience, then what else is it?

Thus, although we have known for a long time that what Bethe said about energy production in the Sun is true, the first real proof that nuclear reactions are indeed taking place in the core of the Sun has come to us from this experiment of Davis. As has been pointed out elsewhere, Davis not only detected the neutrinos emitted by the Sun, but as a bonus also gave us the Solar Neutrino Problem! That has been discussed in the main body of this chapter. Naturally, Davis was rewarded for his marathon effort with the Nobel Prize in 2002. He became a Laureate at the age of 87!

to be more precise, can be produced in various reactions and with various energies. The neutrino energy spectrum is sketched in Fig. 11.4. As you know, the neutrino was detected for the first time in 1956 (see Volume 1). Subsequently, many efforts have been mounted to detect solar neutrinos using detectors of various types. Basically, one makes use of the reaction that the neutrino can cause when interacting with a chlorine nucleus or a gallium nucleus or just water. Experiments are now in progress using all the types of detectors. As can be seen from Fig. 11.4, the different types have

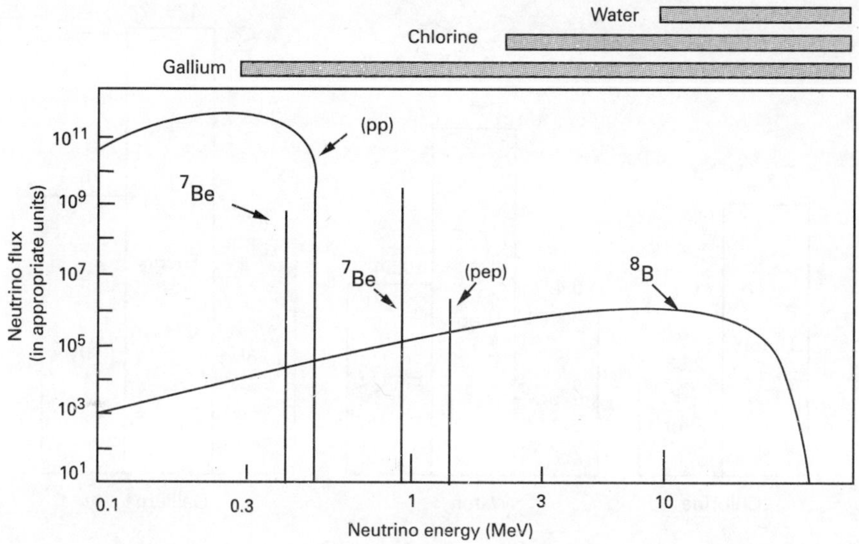

Fig 11.4 *In BOX 11.1 are given the principal nuclear reactions responsible for energy production in the Sun. In many of these, neutrinos are emitted. Shown above are the neutrino energy spectra for the various reactions. In some cases the spectrum is discrete while in others it is continuous. Also indicated by the bands are the energy ranges over which detectors using gallium, chlorine and water are effective. All the three types of detectors have been employed for detecting solar neutrinos.*

different thresholds for detection. While the gallium detector can detect neutrinos having energies about 0.2 MeV and above, the water-based detector is sensitive only to high-energy neutrinos.

Figure 11.5 shows the measured and calculated event rates in five experiments that have been running. Historically, the chlorine experiment due to Davis is the oldest – see BOX 11.2. For years it was the only experiment that gave any information at all about neutrinos coming from the Sun. Only later were the other experiments set up, partly in order to check out the rather puzzling result obtained in the Davis experiment.

Is there any way of estimating what is the neutrino flux from the Sun? Yes there is. The rate at which the Sun emits neutrinos is calculated with the help of what is called the Standard Solar Model (SSM). The model starts off with reasonable assumptions. It assumes that the interior of the Sun is in hydrostatic equilibrium maintained by the balance of gravitational force and radiation pressure. Nuclear fusion as described by Bethe is supposed to be the source of energy. Energy is transported in the Sun mainly by photons and convective currents. And finally, the nuclear abundances are supposed to be changed solely by nuclear reactions. All these are quite reasonable assumptions.

Let us get back to Fig. 11.5. The results are quoted in terms of SNU or solar neutrino units, which is a convenient unit for describing the results of such experiments. We need not be bothered about the details of SNU. What is more

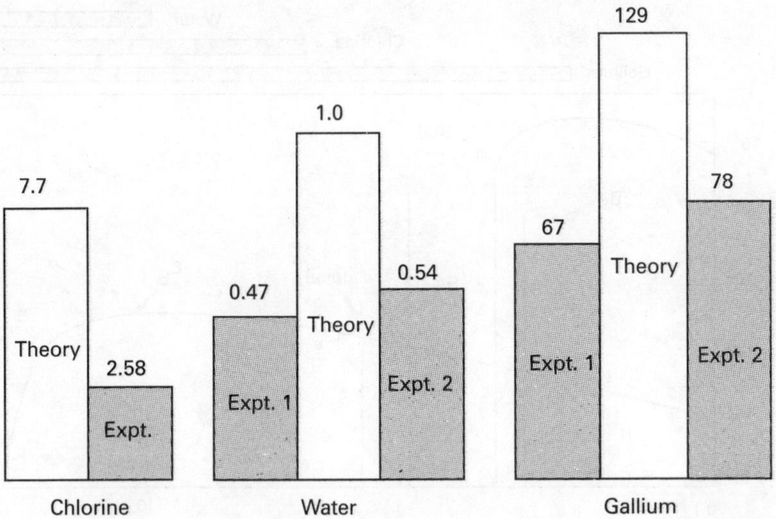

Fig 11.5 *This figure offers a comparison of the observed solar neutrino fluxes and the theoretical estimates of the same based on the SSM. There are five experiments, and their results are shown in shaded bars. To keep the diagram simple, the error margins are not shown. The theoretical estimates are shown via unshaded bars. As mentioned in the main text, the results are quoted in terms of SNU. What stands out is that in every experiment, what is measured is much less than what is predicted, well within experimental uncertainty. Where have the missing neutrinos gone? That in essence is the solar neutrino problem.*

important for us is the following: In every case, the number of neutrinos actually observed is less than what is predicted theoretically based on the SSM, and with a bit of neutrino physics based on the Standard Model of particle physics.

11.5 Experimental evidence for neutrino oscillations

When experiments to detect neutrinos from the Sun were first started a long time, neutrino oscillations were not in the limelight. However, when the solar neutrino puzzle came to the fore, the neutrino oscillation issue began to receive serious attention. People started asking: "Are we seeing fewer neutrinos because of oscillations?" One thing led to another, and soon the huge water detectors set up for detecting proton decay were redeployed specifically for detecting the oscillations. That story will come up soon, but before that, may be I should clear one important point.

You might wonder: "How do we know for sure that the neutrinos that these people involved in detecting are actually from the Sun and not from somewhere else?" This is a very pertinent doubt because neutrinos can hit the detector thanks to cosmic ray interactions in the upper atmosphere, or from objects in the Cosmos other than the Sun. Jumping the story a bit, let me draw your attention to Fig. 11.6. This data was published by a Japanese group (whose experiments I shall describe presently). In the setup used by this group, it was possible to determine the *direction* from which the neutrino came. As can be seen, the peak at zero degrees clearly shows that the neutrinos selected for analysis came from the Sun. The detector was thus like a neutrino telescope really.

OK, less neutrinos from the Sun are seen than one ought to, and we have a mystery, *The Case of the Missing Neutrino*, if you are a Perry Mason fan. Two things could possibly explain the mystery. First, the Standard Solar Model could be wrong, and second, the Standard Model of particle physics could be wrong. Which of these two models is the culprit? That is what I shall describe next.

Historically, the solar neutrino experiment was started with a view to obtaining some information about the nuclear processes taking place in the interior of the Sun. The idea was: If we know in detail what is happening in the star that is closest to us namely the Sun, then we can understand stellar evolution in general better and thereby have a better idea of galaxies, etc. In other words, solar neutrinos were *originally* hunted for as messengers from the Sun. But thanks to the puzzle, the focus shifted, for at least a decade or more, from the Sun to the basic nature of the neutrino itself. Could the so-called solar neutrino puzzle tell us something about the neutrino that we do not know for sure yet? In fact, it has turned out that way. By the way, it is interesting to note in passing that the neutrino once born leaves the Sun almost immediately and begins its journey in space. And remember that the neutrino

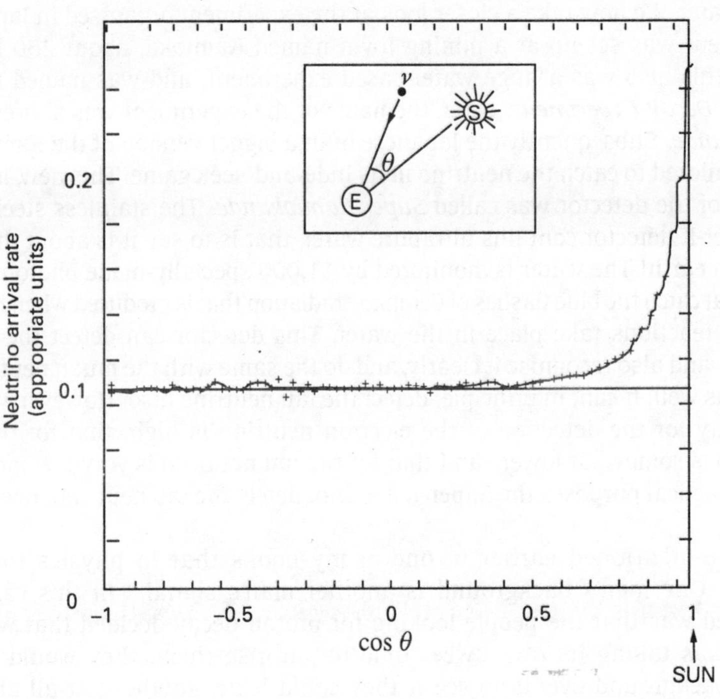

Fig. 11.6 *This figure shows the neutrino data collected over a period of 1258 days by Japanese scientists. What is plotted is the arrival rate vs. cos θ, where θ is the angle with respect to the Sun (see inset). The peak at cos θ = 1 (implying θ = 0) is due to solar neutrinos. What it means is that the detector allows one to select events that one can attribute safely to neutrinos that have come from the Sun.*

is born deep within the Sun, right near the core where the temperature is high-enough for sustaining thermonuclear reactions. On the other hand, the photons that we see coming from the Sun emerge from the *surface*, and they might have spent even a million years inside the Sun, getting scattered here and there, before escaping!

Once the solar neutrino puzzle came to the fore, people examined both possibilities —that the Standard Solar Model (SSM) of solar physics is wrong or that the Standard Model (SM) of particle physics is inadequate. Careful examination revealed that the needle of suspicion pointed towards the latter. It was clear that the SM was not enough; the neutrino seemed to be doing something that was not expected within the framework of the SM. What was that? Oscillations of course. Neutrino oscillations have been talked about for quite some time and the focus now shifted to that. In short, everybody wanted the 'smoking gun', a clear proof that there were in fact neutrino oscillations. This takes us on to the interesting part of the neutrino story, namely the oscillations. To wrap up what was said so far, in a manner of speaking, the solar neutrino puzzle strengthened the case for neutrino oscillations, and induced physicists to look for *direct* experimental evidence.

You will recall that many proton-decay experiments were set up when the GUT models came along and everyone was excitedly proclaiming that the proton could decay. A brief reference was made in Chapter 3 to two such experiments using water. We now take a closer look at the experiment organised in Japan. This experiment was set up at a mining town named Kamioka, about 250 km from Tokyo. This also was a huge water-based experiment, and was named *Kamioka Nucleon Decay Experiment*. Later, the name of the experiment was abbreviated to *Kamiokande.* Subsequently the Japanese built a bigger version of the Kamiokande setup, tailored to catch the neutrino in its hide-and-seek game. This new, improved version of the detector was called *Super Kamiokande.* The stainless steel tank of the Super-K detector contains ultrapure water, that is to say it is about the purest water on earth! The water is monitored by 11,000 specially-made photomultiplier tubes that catch the blue flashes of Cerenkov radiation that is produced when neutrino-induced reactions take place in the water. This detector can detect the electron neutrino and also recognise it clearly, and do the same with the muon neutrino and muons as well. It can, in principle, detect the tau neutrino also. However, while the sensitivity for the detection of the electron neutrino is high, that for the muon neutrino is somewhat lower, and that for the tau neutrino is very low indeed; so, for all practical purposes the Super-K does not detect the tau neutrino. Keep that in mind.

I have mentioned earlier in one of my books that in physics there is a saying: "One man's background is another man's signal." In this case what happened was that the people looking for proton decay decided that while the proton was taking its own sweet time to surprise them, they would analyse all the background events to see if they could learn anything at all about the background. The most important source of background was the neutrino produced in the atmosphere by cosmic rays impinging from outside.

By the way, let me remind you once more how elusive these neutrinos are so that you develop admiration for the experiments that I shall be describing later. Every second, about 60 billion neutrinos pass through every square cm of our body, and without doing anything! In fact, it is said that one could send a beam of such neutrinos through one light year of lead, and most of the neutrinos would come out safe at the other end! OK, now when these scientists waiting for the proton decay started analysing the neutrino background, they found faint evidence of neutrino oscillations. Naturally they became excited and said, "Hey, why not focus on catching neutrino oscillations? That way, we would be getting at least some positive results from these monster apparatus on which we have spent so much money and energy." And that is when the ball game changed from detecting proton decay to hunting for neutrino oscillations. If the neutrino was really oscillating, then it also meant that the neutrino had mass. So people said: "Can we catch the neutrino oscillating and can we place stringent limits on the neutrino mass?"

OK, now what kind of neutrinos hit the detector? We know for sure that neutrinos could come from the atmosphere thanks to cosmic rays, and they also come from the Sun. For the moment, we leave out other possibilities. Let us now see how the Super-K scientists used atmospheric neutrinos, long considered a damn nuisance, to demonstrate that oscillations were occurring.

Take a good look first at the first of the following equations and then at Fig. 11.7. When cosmic rays hit the nuclei in the upper atmosphere, they produce a lot of pions. Pions, as you know, decay into muons, and later the muons themselves decay into electrons. The relevant decay equations are

$$\pi^+ \rightarrow \mu^+ + \nu_\mu , \tag{7a}$$

$$\mu^+ \rightarrow e^+ + \nu_\mu + \nu_e . \tag{7b}$$

We must of course not forget the charge conjugate equations like

$$\pi^- \rightarrow \mu^- + \bar{\nu}_\mu , \text{ etc.}$$

So from the above we see that thanks to the π–μ decay, for every such chain, there must be two muon neutrinos and one electron neutrino.

Suppose there were no neutrino oscillations. Then both the electron and muon neutrinos would retain their respective identities, and if one could detect both types of neutrinos, then one should find the ratio of muon to electron neutrino to be 2:1. The scientists at Super-K decided to check this out. Now to perform this experiment, the detector ought to be able to clearly distinguish between the muon and the electron neutrinos; this the Super-K detector could do. Preliminary results showed that the ratio was less than 2; it seemed that oscillations were indeed taking place. At this stage, the detectives, if I might call the scientists at Super-K that, made their experiment more subtle and also interesting. I want you to follow this carefully so that you can appreciate the ingenuity and the imagination of the experimenters doing such complex experiments. This is where a reference again to Fig. 11.7 becomes necessary. This figure tells us that the Super-K detector is able to say from which direction the neutrino came. This is important.

Bearing all this in mind, now look at Fig. 11.8. This shows the number of muon neutrinos detected by the Super-K as a function of the angle theta, or

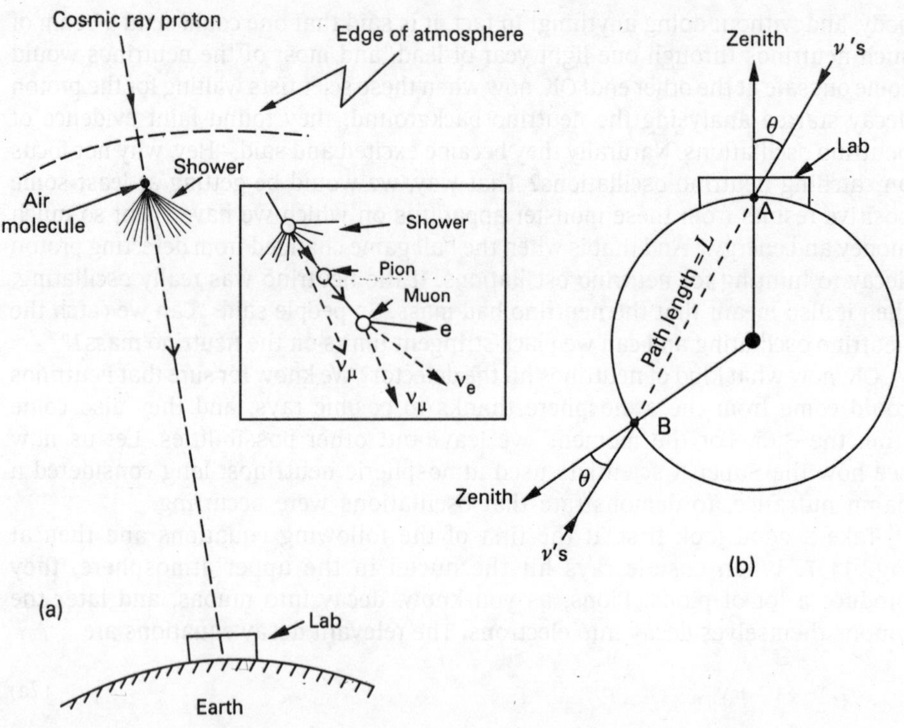

Fig. 11.7 *In this figure, (a) shows schematically the generation of the two types of neutrinos in the upper atmosphere by the incoming cosmic rays [according to equation (7)]. Notice that for every electron neutrino, there are two muon neutrinos. These neutrinos are usually produced several kilometres above the Earth and since the pion and the muon would be travelling at very high velocities, the directions of the three neutrinos produced in a given event would be more or less the same. (b) shows something interesting. A and B are two points on the surface of the Earth, with the neutrino lab being at A. Consider neutrinos entering the lab at an angle θ with respect to the zenith. Neutrinos with a similar angle with respect to the zenith may strike B. And these neutrinos could travel through the Earth from B to reach A. However, unlike the neutrinos coming directly to A, those coming from B would have time to oscillate. Thus, whereas the detector would detect two muon neutrinos if the hit is from above, it would register a smaller number if the muon neutrinos come 'below', that is, from B. This is a very important point. Now the Super-K could distinguish clearly the direction from which the neutrinos came, i.e., whether they came direct to A or from B and at what angle. It is this feature that helped in deducing the data presented in the next figure.*

equivalently, the distance travelled. Remember that oscillation would convert the muon neutrino into a tau neutrino, which would go undetected by the Super-K. The scientists of Super-K compared their results with what might be expected if there were no oscillations and if there were oscillations. As you can see from Fig. 11.8, the results leave no room for any doubt—neutrino oscillations are a reality.

Let us pause and take stock. It all started with Davis trying to catch neutrinos from the Sun and laying the foundations for neutrino astronomy. But he detected

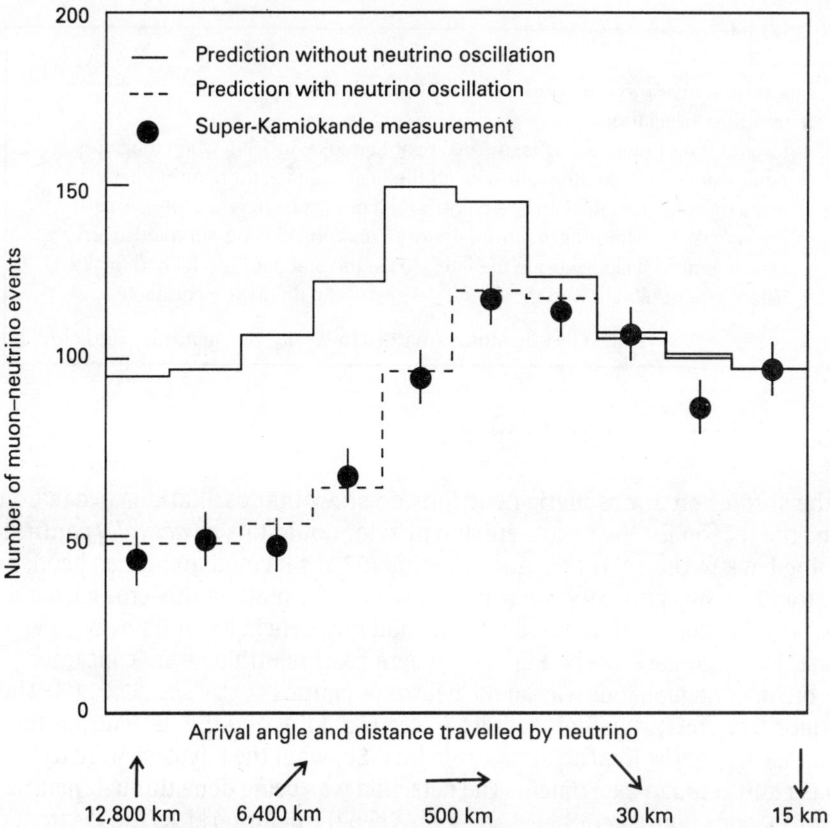

Fig. 11.8 *This figure shows the number of muon neutrinos arriving at the Super-K detector from different directions and after travelling through various distances in the Earth (as described in the previous figure). If there were no neutrino oscillations, then the distribution expected is that given by the top curve with a full line. The observed data however, fall much below, especially when the incoming muon neutrino has to travel a long distance through the earth. During the travel, many muon neutrinos become tau neutrinos and escape detection. Thus, the number of events observed with 12,800 km of travel is almost half of that which is expected without oscillation. This result provided the first indisputable evidence for neutrino oscillation.*

fewer neutrinos that the Standard Solar Model (SSM) predicted. That meant that either the SSM was defective or the Standard Model of particle physics used to estimate the counting rate in the detector was defective. People believed that neutrino oscillation might explain the discrepancy and thereafter the Super-K experiment on atmospheric neutrinos (reported in 1998) did strongly suggest, as Fig. 11.8 shows, that neutrino oscillation was indeed a real phenomenon. See also BOX 11.3.

BOX 11.3

This is how an American professor recalls the announcement of the discovery of neutrino oscillations:

> I was at the Conference at Takoyama, near Kamioka, in 1998 when the Super Kamiokande collaboration announced the first evidence for neutrino mass. It was a moving moment. Uncharacteristic for a physics conference, people gave the speaker a standing ovation. I stood up too. Having survived every experimental challenge since the late 1970s, the Standard Model had finally fallen. The results showed that at the very least, the theory is incomplete.

Prof. Hitoshi Murayama of University of California, Berkeley

11.6 At last the smoking gun

OK, the studies on atmospheric neutrinos do show that oscillations occur. Could this be the reason for the solar neutrino puzzle? Could this be *actually* confirmed by going back to the solar neutrino experiment? Yes it could and it has been, and that story has two parts. By the way, one reason for making this crosscheck was to see whether the Standard Solar Model had any deficiency of its own.

First, let us go back to the Super-K. Where solar neutrinos were concerned, the focus of Super Kamiokande was on the neutrinos emitted by ^8B – see BOX 11.1. These neutrinos have relatively high energies – see Fig. 11.4, and the ^8B neutrino flux is only about 10^{-4} of the total solar neutrino flux. So, what the Japanese were trying to do was really a tough experiment. The detection was being done through neutrino–electron elastic scattering (ES), i.e., $\nu_e \rightarrow \nu_e$. When the neutrino kicks the electron, the recoiling electron generates Cerenkov radiation, which is what is picked up by the photomultiplier bank. After patiently collecting data for 1258 days, the Super-K scientists found that the ^8B neutrino flux detected was definitely less than what the Standard Solar Model predicted. You might say, "So what? This was already known, even from the chlorine experiment of Davies." True, but be patient, for this is where the story gets real interesting!

Enter now a new star player, the SNO. SNO stands for Sudbury Neutrino Observatory. It is located two kilometres below ground in a mine about 300 km north of Toronto in Canada. Now Canada is a very large country, almost as big as the United States in area. However, its population is just about a tenth. It is natural therefore that Canada cannot quite compete with the US in science. However, the Canadians are smart. They know that they cannot match the US in sheer quantity, but they try to make up by selecting a few areas for research and making every effort to be the best in those chosen areas. It is this philosophy that led to the creation of the SNO.

The Canadians said, "We cannot spend as much money as the US. However, let us set up one experiment that would really produce a world-class result." Arguing in this manner, they voted for neutrino physics. Now Canada had a large heavy-water reactor power programme (like India – see ***Bhabha***). They thus had many plants that produce heavy water, that is, D_2O. Heavy water is quite expensive but this was a matter of prestige and world-class excellence; so the Canadian Government agreed to give 1000 tons of D_2O. This heavy water is contained in a spherical vessel 12 metres in diameter. In turn, this spherical vessel is immersed in a 30 metre cavity filled with ordinary water. The interaction of the neutrino with the deuteron in D_2O is detected with a bank of 10,000 photomultipliers.

This brings us to a point we have not discussed so far. How do the neutrinos interact with the atoms in the detector? In the case of ordinary water, the neutrino bounces off one of the electrons in the water molecule, and this is the kick given to the electron that produces the Cerenkov flash. In other words, the neutrino is detected as a result of the elastic scattering (ES) that occurs. By the way, although in principle all neutrino types can cause ES, ES is weak in the case of the muon and the tau variety of neutrinos.Hence, with a water detector one monitors mainly the electron neutrinos. When it comes to heavy water, we have three possibilities as below:

$$\nu_x + e^- \rightarrow \nu_x + e^- \text{ (ES)} \qquad [x = e, \mu, \tau]$$

$$\nu_e + d \rightarrow p + p + e^- \text{ (CC)}$$

$$\nu_x + d \rightarrow p + n + \nu_x \text{ (NC)} \qquad [x = e, \mu, \tau].$$

In other words, apart from detecting elastic scattering (ES) which is also possible in the Super-K detector, the SNO detector can detect neutrinos via the charge current (CC) and neutral current (NC) processes also (see Volume 1, Chapter 10 for a description of CC and NC processes). Remember that ES is basically sensitive only to ν_e , though in principle this process can occur for all the three flavours of neutrinos. In contrast, the NC process is *equally* sensitive to all flavours. This is the trump card that the Canadians played, by going in for heavy water while designing their detector. See also Fig. 11.9.

So what happened? Did the Canadians get the pay off? They sure did! In July 2002, they published their epoch-making result. Like the Japanese, they too were catching the ^{8}B neutrinos. But while the Japanese relied on ES which mostly misses the mu and the tau neutrinos due to poor sensitivity, the SNO experimenters had all the avenues open to them. They could selectively catch the ν_e's from the ^{8}B reaction in the Sun via ES; they could also catch the mu and the tau neutrinos born from the original ν_e's via oscillation through the NC process as well as the remnant ν_e's.

Let us now look at what the SNO team finally reported. (They gave their results in actual flux instead of SNUs, and I am quoting what they reported, leaving out their error estimates.) First, take a look at Fig. 11.10. This shows the space of the electron neutrino flux and the combined μ and τ neutrino flux. The data collection was for 306 days, and the bands in this figure show the regions of this space consistent with the three different types of observations—ES, CC and NC. By the way, the ES result of SNO agreed very well with the ES result earlier reported by the Super-K. So, that means that the SNO observations are very credible. Notice that the bands all intersect in a region.

OK, so what does it all mean in terms of the so-called solar neutrino puzzle? Well, the SNO reports a total ^{8}B neutrino flux of 6.42 in the units used in Fig. 11.10.

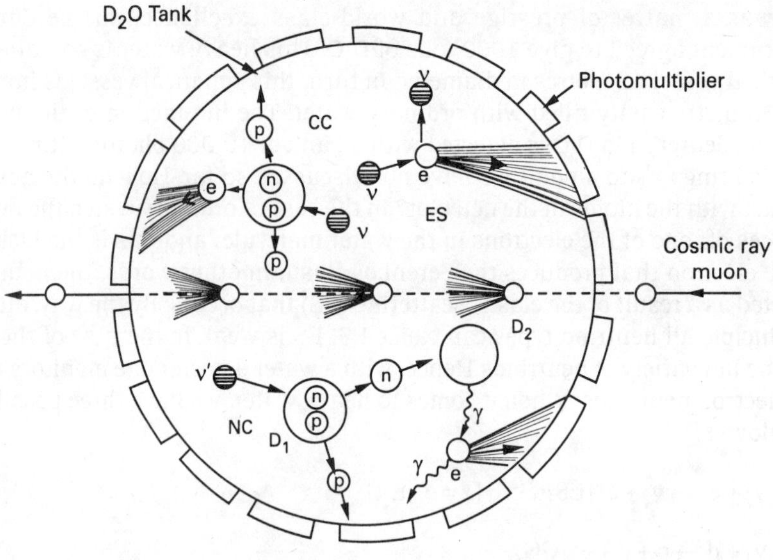

Fig. 11.9 *This figure illustrates schematically how neutrinos are detected in the SNO detector system. As explained in the text, the detection occurs via three distinct mechanisms – ES, CC and NC processes. In each case, visible photons are ultimately produced, and it is when these are picked up by the photomultiplier bank that detection finally occurs. In ES an electron is kicked by the incoming neutrino, and it is this recoiling electron that generates the Cerenkov flash. In CC the neutrino is absorbed by the neutron in the deuteron, producing a proton and an electron. The electron causes the Cerenkov flash. And in NC the neutrino splits the deuteron, knocking out the neutron. This neutron is then captured later by another deuterium (^2H) nucleus to become a tritium (^3H) nucleus. A gamma ray is produced in the process, which then scatters electrons to produce visible flashes. Cosmic ray muons can be recognised because they pass through straight emitting Cerenkov radiation all through their transit. This is the way background signals are rejected.*

This is the measured value and comes from the NC reaction which is sensitive to *all* types of neutrinos. For comparison, the Standard Solar Model gives a value total ^8B flux of 5.05. If one considers the errors attached to both theory and experiment, then the agreement is quite good.

What is the implication of this? Let us go back to the solar neutrino problem. Suppose the Sun emitted only electron-type neutrinos and there was no such thing as neutrino oscillation. Then both the Super-K and SNO would have detected the same neutrino flux. But what actually happened was that the Super-K (which relies on the ES process) detected less. The SNO too saw a lower flux when detection was by the ES process alone. However, when the detection was by the NC process, all the neutrinos that had changed flavour and had become μ or τ neutrino were also caught (in addition to the ν_e's), and there was now no question of any missing neutrinos. This total number was, within the limits of error, the same as that predicted by the Standard Solar Model. As the SNO scientists said in their paper, "The total flux of ^8B neutrinos measured with the NC reaction is in agreement with the SSM

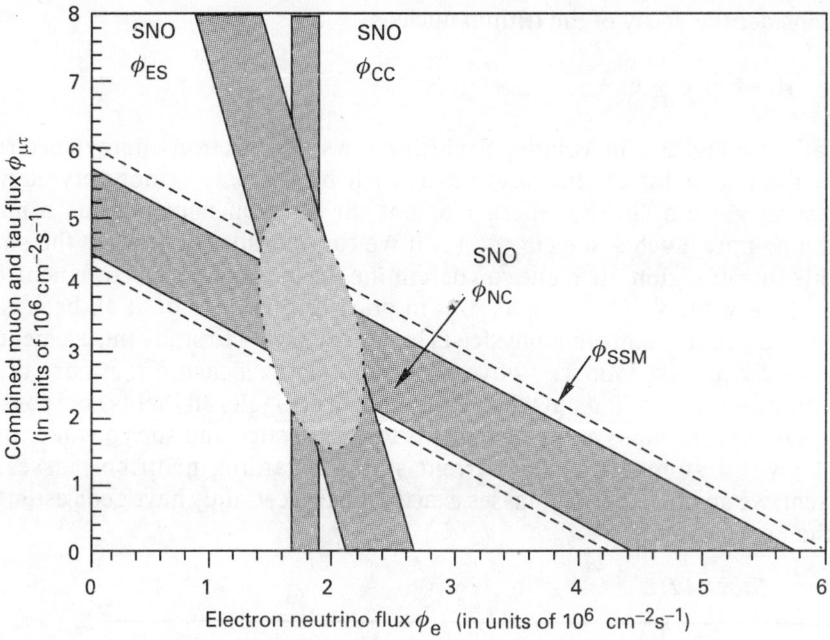

Fig. 11.10 *This figure presents a map of the space of neutrino fluxes from the Sun associated with ^{8}B decay. The neutrinos are born as electron neutrinos, but during their transit to Earth they could transform first into the muon type and later into the tau type. The x-axis shows the electron neutrino flux ϕ_e, while the y-axis shows the combined muon and tau neutrino fluxes, i.e., $\phi_{\mu\tau}$. The three bands show the ranges indicated by the various measurements made by SNO in this space, bounded by the two axes as shown. The results for the Standard Solar Model in this space are also indicated. If one reads off the fluxes at the intersection point of ES, CC and NC, one finds that $\phi_e + \phi_{\mu\tau}$ is equal to the total flux of electron neutrinos predicted by the SSM, within experimental errors of course. This is the long-awaited smoking gun.*

prediction." In other words, the SNO really clinched the case for neutrino oscillations, using solar neutrinos—this was hailed as the smoking gun!

So at last, thanks to SNO and of course the pioneer Super-K, one can now say with absolute certainty that (1) there *is* such a thing as neutrino oscillation, (2) the neutrino mass is definitely *not* zero, (3) one *has* to go beyond the Standard Model of elementary particles, and (4) the Standard Solar Model is quite robust.

11.7 Mass of the neutrino – present status

If neutrino oscillations are real, it automatically means that the neutrino *does* have a non-zero mass. OK, agreed that the mass of the neutrino can no longer be taken to be zero; but what exactly is the value of the mass? What do the experiments say? The cleanest way of determining the mass of the electron neutrino is to go back to good old-fashioned beta decay.

Consider the decay of the tritium nucleus:

$$^3H \rightarrow {}^3He + e^- + \bar{\nu}_e \, .$$

Recall now Fig. 5.1 in Volume 1 which shows the electron energy spectrum. Now the end point of the spectrum (which by the way is not very easy to measure) gives a clue to whether or not the electron neutrino has a mass, and if so how much – see Fig. 11.11. If we can measure accurately the shape of this cut-off region, then one can determine the mass of the electron neutrino. It is all very simple in principle. But in practice this has not at all been easy. For decades, very competent physicists have been trying to study this elementary beta decay process. Yet they have not been able to measure the mass of the neutrino accurately. Find out why! After years of struggle, all that is now possible is to say that the mass of the neutrino is less than such and such a value. Table 11.1 gives a summary of the current status regarding neutrino masses. At present, we do not know the masses exactly, but we certainly have good estimates of the upper bound.

Table 11.1

Type	Mass estimate
Electron neutrino	< 2.2 eV
Muon neutrino	< 170 keV
Tau neutrino	< 18.2 MeV

11.8 Neutrino helicity and its non-zero mass

We now come to an important point. For decades we generally accepted that the neutrino mass was really zero, though theorists were cautious and experimenters kept on saying that experimental evidence for the mass being zero is lacking. And then came the two-component neutrino theory and all that, which was discussed in Volume 1 (see Chapter 5). Closely related to this is the concept of *helicity*, and I also described a great experiment due to Goldhaber that showed the neutrino to be left-handed. In turn this implied that the neutrino mass was zero; otherwise it could not have a helicity. What happens to all that now?

Now the electron, which is a particle with non-zero mass, is described by Dirac equation and can be viewed in one of the two possible ways. One way is the familiar spin-up and spin-down picture, transverse to the direction of motion of the electron. The other is through *chirality*, that is, spin projected parallel to the direction of motion (right-handed), or antiparallel to it (left-handed), depending on its velocity. If it is right-handed in one Lorentz frame, we can always find a frame in which the direction of the momentum is reversed and appears left-handed. *This is not possible for a massless particle.* However, in the case of a particle with small mass, though in principle there are two chirality states, one

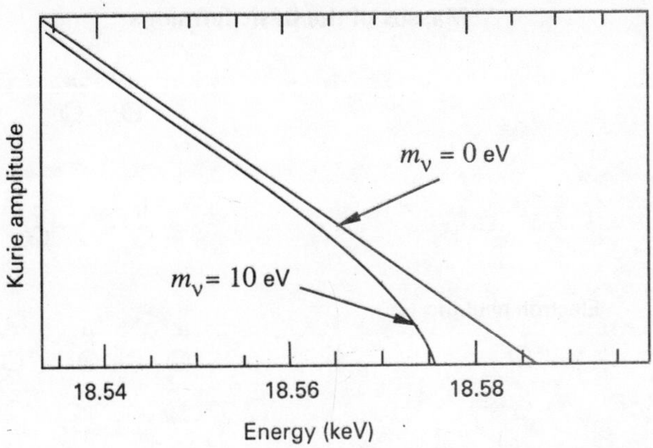

Fig. 11.11 *This figure shows qualitatively how the end point of the beta spectrum is analysed to extract information about the neutrino mass. As mentioned in the text, the simplest beta decay to study is that of tritium. The measured electron energy spectrum is then replotted as what is called a Kurie plot. As the figure above shows, the end point of such a plot is very sensitive to the mass of the neutrino. So, it all boils down to measuring the end point of the electron energy spectrum very accurately. Amazing that an apparently simple experiment like this is presenting so many problems, when people are able to look for neutrinos from the Sun and the stars! Sometimes, that is the way it is!*

chirality state (or state of handedness) will dominate over the other. In the limit $m \rightarrow 0$, the dominant chirality state will become the *helicity* state.

When I earlier described the Goldhaber experiment (see Volume 1, chapter 5), I stated that he had determined the neutrino to be left-handed possessing a unique helicity. This was OK at that time because everyone believed that the neutrino had zero mass. In the context of the discovery of neutrino oscillations which means that the neutrinos have mass (although very small), the conclusions of the Goldhaber experiment have to be re-examined. When this is done we find that Goldhaber *actually measured the dominant chirality state for the small mass neutrino*, and within experimental errors found it to be left-handed. Neutrinos with the opposite chirality are much fewer in number and did not get detected. I hope that clarifies what the revised interpretation is.

11.9 Next generation neutrino experiments

To be fair, as mentioned earlier, for quite some time people have been looking beyond the Standard Model. Once one does that, one is ready for neutrino oscillations and non-zero mass. In fact, the oscillations did not at all come as a surprise; rather, they were looked for and found. But questions do remain. First, how does the mass *actually* arise? Next, why is the neutrino mass so low compared to the masses of the other fermions – see Fig. 11.12. Thirdly, is the neutrino a Dirac particle or a Majorana particle? Who says neutrino physics is not exciting?!

What next? For one thing, there is a move to study accurately the difference in the neutrino reaction in Super-K between day and night. The neutrinos arriving in

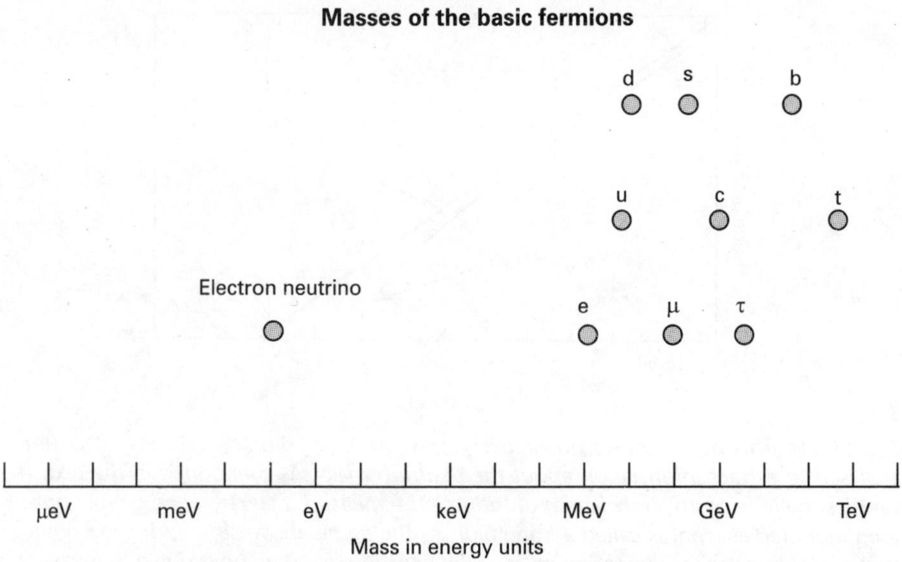

Fig. 11.12 *A comparison of the masses of the fundamental fermions, with mass indicated in energy units. Notice how light the electron neutrino is compared to the electron. At present we do not have exact numbers for the neutrino masses, but it is certain that in general their masses are way below that of the correspnding leptons. Question: How do neutrinos acquire mass, and why are they so light compared to the associated lepton? Future will tell!*

the detector at night have to travel through the Earth, i.e., along a longer path, than those arriving during the day. So, one should be able to get some more data on the oscillations. See Fig. 11.13. On the other hand, there is also a move to stop depending on free but uncontrolled neutrinos supplied by cosmic rays, and switch to 'paid' neutrinos, that is, muon neutrinos produced in accelerators under controlled conditions. Two big programmes are being planned, one in Japan and the other in America. In Japan, it is planned to create a neutrino beam in the KEK accelerator laboratory which is about 250 km away from the Super-K detector. The beam can be turned off and on as desired, and it would also have a well-defined energy as well as direction. Moreover, it is planned to have a replica of Super-K right next to the source. This would help to characterise the muon neutrinos before they oscillate. A similar experiment named MINOS is planned in the US, where neutrinos sent from Fermilab would travel 730 km and be detected by a detector that is being built in Minnesota. This experiment hopes to study the transformation of muon neutrinos produced in Fermilab into τ neutrinos via oscillations while travelling the distance of 730 km. Similar experiments are planned in Europe also, based on muons produced in CERN.

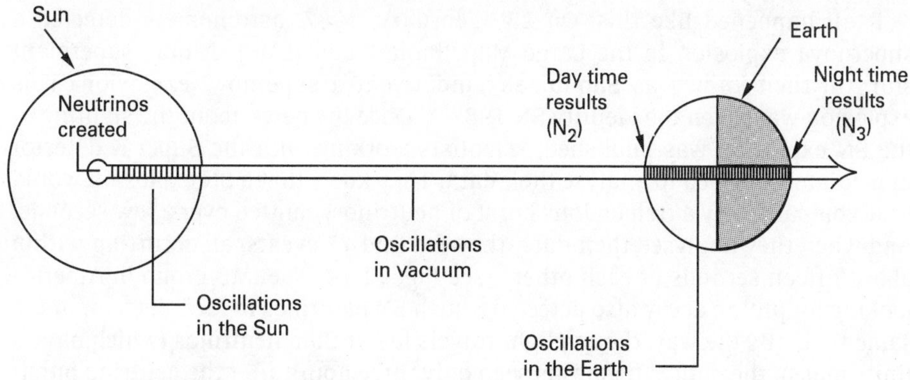

Fig. 11.13 *This figure illustrates schematically the various stages of the oscillations that take place while neutrinos travel from the Sun to the Earth. Oscillations become enhanced when there is matter. If N_1 electron neutrinos leave the centre of the Sun on their way to the Earth, then due to oscillations only a lesser number N_2 of the electron neutrinos will reach a detector on Earth when it is directly facing the Sun, i.e., during day. At night, due to the transit through the Earth, the number reaching would come down to N_3. This is the day-night effect.*

11.10 And now, neutrino astronomy!

From solar neutrinos we turn to another cosmic source for these particles. On 23rd February, 1987, something very remarkable happened. On that day, a supernova explosion was observed in the sky. It was the first supernova visible to the naked eye since Kepler's supernova in 1604. That, however, is not the remarkable part. On that day, two experiments looking for proton decay both detected neutrinos from the supernova. Just to remind you, the supernova explosion occurs when the electron–proton plasma in the star (with mass exceeding the Chandrasekhar limit – see ***Chandra***) gets crushed by gravity to become a neutron fluid according to the equation

$$e^- + p \rightarrow n + \nu.$$

This neutrino burst occurs in the core, and it is after this a neutron star is born. A tremendous amount of energy is released, the mean neutrino energy being about 15 MeV. By various other mechanisms, the remainder of the parent star gets excited, and as a consequence electromagnetic radiation is also produced. It is as a result of this latter process that there is a bright flash in the sky, and that is the phenomenon that Kepler saw in 1604, and the Chinese in 1054 (see ***Chandra***). Turning to the energy release in the SN explosion, it is estimated that it was about 3×10^{53} ergs in the supernova explosion seen on 23rd February 1987. Two-thirds of this energy was carried by neutrinos and roughly one third by anti-neutrinos that are also produced during the explosion. About 10^{58} neutrinos and anti-neutrinos (together) are estimated to have been emitted in the supernova explosion referred to above. Of these, about 10^{16} particles hit the Super-K detector, and of these only twelve, yes twelve, were detected! Yet, how much they revealed and what a stir they created!

It all happened like this. On 23rd February, 1987, astronomers detected a supernova explosion in the Large Magellanic Cloud (LMC). A blue supergiant star, till then known as Sanduleak, underwent a supernova explosion. This explosion was given the identity SN 1987 A. Once the news about the sighting of the SN explosion was published, scientists working with the Super-K detector immediately decided to analyse their data. They knew that a SN explosion would be accompanied by a tremendous burst of neutrinos, emitted over a few seconds. And when they analysed their data, they did find 12 events, all occurring within about fifteen seconds of each other – see Fig. 11.14. The IMB group in America looking for proton decay also detected 8 such SN neutrinos in their detector at the same time. (By the way, though light travels faster than neutrinos (which have a finite mass), the optical flash was seen only three hours after the neutrino burst! Any ideas about why the optical flash arrived later?) I want you to appreciate that the Super-K scientists were *not* looking for neutrinos from supernova explosions. But when they heard that a supernova had been sighted, they immediately went to the computer hooked to the detector and started analysing all the data accumulated around that time. This they could do because the detector was working non-stop, 24 hours a day. And sure enough, when they went back to the data, they found precisely what they were looking for.

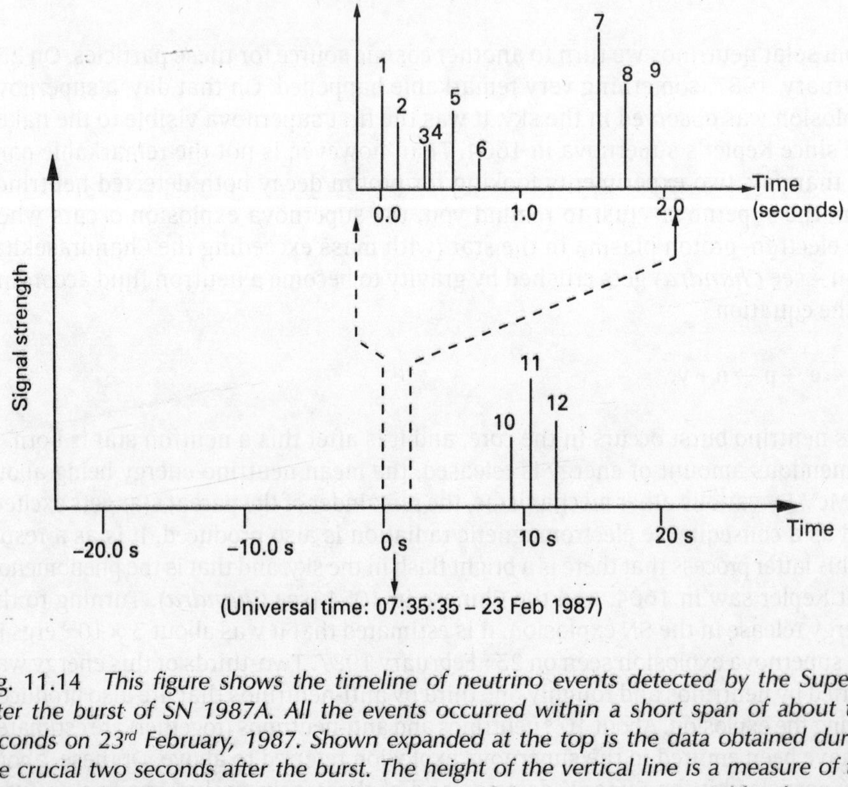

Fig. 11.14 *This figure shows the timeline of neutrino events detected by the Super-K after the burst of SN 1987A. All the events occurred within a short span of about ten seconds on 23rd February, 1987. Shown expanded at the top is the data obtained during the crucial two seconds after the burst. The height of the vertical line is a measure of the strength of the photomultiplier signal. Every one of the 12 events reported has been very carefully checked before acceptance.*

BOX 11.4

Thus far we have had, besides good old optical astronomy, other types such as radio astronomy, infra-red astronomy, X-ray astronomy, and gamma ray astronomy, all based on the electromagnetic spectrum. Presently, physicists are daring to use that most elusive particle, the neutrino, to launch neutrino astronomy. The basic idea is to detect high-energy neutrinos coming from outside and in this manner gain new information about some of the exotic objects in the cosmos.

One system that is planned in this connection is the neutrino telescope named DUMOND, which is an acronym for Deep Underwater Muon and Neutrino Detector. The medium is not a large tank of water as in the Kamkiokande detector, but the sea itself! It has been observed that deep seawater off the islands of Hawaii is unusually transparent. At a depth of below 4.5 kilometres, a flash of blue light is able to travel quite successfully over a distance of 25 metres. The DUMOND has been designed based on this fortunate circumstance.

The DUMOND detector will start off with over a thousand light detectors distributed over a volume of one cubic kilometre of seawater. There would be long cables anchored to the bottom of the ocean. To each cable would be attached many photomultipliers, at regular spacing. There would be an array of such cables, thus distributing light sensors over a large volume of sea water.

The DUMOND array is being designed to look for very high energy neutrinos, that is, neutrinos with energies of 10^9 eV or even as high as 10^{11} and 10^{12} eV. Such high-energy neutrinos are expected to be produced in the cosmos when relativistic protons collide with nuclei. Such collisions could occur in the inter-galactic clouds, at the edge of galaxies and at the boundaries of quasars —there are all kinds of possibilities.

When a high-energy neutrino interacts with a particle in an atom of water molecule in the ocean, it would give rise to a cascade of electrically charged particles, about a few millions of them. These particles will generate Cerenkov radiation, and it is the light flashes generated by the Cerenkov radiation that the detector array must detect.

What about background signals? Well, muons can trigger the neutrino detector if they can reach it. A lot of muons are generated in the upper atmosphere by the cosmic rays (see ***Bhabha*** and also Fig. 11.7), and it is to keep the muon background low that the detector system is being placed almost at the bottom of the deepest ocean in the world.

Other than muons, are there any other sources of background? Yes, the neutrinos produced in the upper atmosphere. We have already discussed these in the main body of the chapter. These neutrinos produced by cosmic rays in the atmosphere could create an unwanted signal. However, by using clever methods , the detector signals due to these could be identified and the background signal rejected.

So what would DUMOND look at, sitting deep under the ocean? It would basically try to catch neutrinos expected to come from exotic sources in the cosmos as already explained. Thus in principle, DUMOND could discover all sorts of suspected and even unsuspected neutrino sources in the cosmos.

Where astronomy is concerned, we must always remember the story of

...contd.

BOX 11.4

contd. ...

Fraunhofer's discovery (see ***Saha***) that led to a detailed study of our Sun and later other stars and even galaxies. The first steps are always small but in due course—who can say what it would lead to?

By the way, I wonder whether fishes and whales would disturb this detector. They say there are also marine organisms that emit light. This is called bio-fluorescence and I wonder if all these could cause problems. Perhaps not, at those depths. In any case, I am sure the designers would have considered all such possibilities and taken due care!

I must tell you something about how careful scientists are with precious data. The Super-K scientists had data about exactly twelve neutrino events believed to be associated with the supernova explosion. They had to be *absolutely* sure this burst came from SN 1987 A and nowhere else. I can't go into all the technical details, but believe me, the data analysis was a model of thoroughness. In science, people are always ready to jump on your neck and tear your findings to pieces, be it experimental or theoretical! Frontier research is that hot and competitive a game. The fact that the results of Super-K regarding SN 1987 A have been universally accepted is a tribute to the care with which the experimenters analysed the data, distilling every bit of tell-tale evidence.

Just imagine. Barely forty years ago people were struggling to just detect the neutrino. And now they are able to detect neutrinos even from supernova explosions! This remarkable first from Super-K and the IMB experiments has already confirmed that the basic ideas of supernova physics are quite sound. The mood is quite upbeat, and physicists are confidently talking of regular and routine neutrino astronomy!

All that I have described so far has set in motion two distinct trends. On the one hand, there are astrophysicists who want to use the neutrino to learn about stars, supernova, black holes and what have you. In particular, one may expect an intensified effort to measure the energy spectrum of the neutrinos coming from the Sun (recall Fig. 11.4). This is very detailed information, and would go a long way in telling us a lot about the precise nuclear reactions taking place within the Sun. In turn, this would help not only in improving the model we now have for the Sun, but also in sharpening our concepts about stellar evolution. So many things once not even dreamt of now seem possible! On the other hand, the tremendous success achieved in recent years has made people consider various subtle experiments that would give more information about the neutrino itself. There are all kinds of questions relating to neutrino mass, how exactly it originates, the implications for cosmic asymmetry where leptons are concerned, and so forth. Taking all this into account, you should appreciate that though this chapter is a bit long, it is not even exposing the tip of the iceberg!

I wonder if all this would have happened if the decay of the proton had been detected. What do you think? I personally think it is a good blessing that proton decay has eluded us thus far; otherwise, we might have missed many of these goodies since attention might have got focussed elsewhere.

There are a few interesting things I have tucked away in BOXES 11.4 and 11.5. You must at least glance through them for they would give you some idea of the wonderful panorama of neutrino physics today, dealing with aspects that I have not described in the main body of the chapter.

11.11 Summary

- Many attempts are in progress to detect double beta decay. A definitive result is expected soon.
- The neutrino oscillations are for real, and they indicate clearly that there is physics beyond SM.
- The long-standing solar neutrino puzzle has finally been solved. The observed discrepancy is due to neutrino oscillations and not due to any shortcoming of the Standard Solar Model.
- Neutrinos produced in supernova explosions have been detected, paving the way for neutrino astronomy.
- Newer detector systems are being planned. While some are for astronomy, others are for elucidating finer aspects relating to oscillations.
- Exact masses of the neutrinos of different flavours are not yet known, but fairly tight upper limits have been set.

NEUTRINO TIMELINE

1920: Eddington suggests that energy is produced in the Sun through nuclear reactions.

1930: Pauli puts forth the idea of neutrino but calls it neutron.

1933: Fermi names the neutrino and makes the concept firm.

1938: Bethe describes in detail the fusion chain in the Sun and accurately estimates the central temperature.

1956: Reines and Cowan detect the neutrino.

1962: Muon neutrino detected.

1964: John Bahcall estimates the neutrino flux expected from the Sun.

1967: Raymond Davis detects solar neutrinos for the first time.

1969: Gribov and Pontocorvo suggest that neutrino oscillations could explain the solar neutrino deficiency.

BOX 11.5

A long time ago, accelerators produced just a handful of particles. But soon, by clever design, the particle flux was made copious and physicists began to talk of factories. First there was the meson factory, then the B factory and now there is talk of a neutrino factory.

Why a neutrino factory? Physicists say that a neutrino factory would enable them to study neutrino oscillations in *very great detail*. No doubt there are already some proposals to squeeze neutrinos out of some of the big accelerators around (like Fermilab and CERN, for example), but a factory would produce a much more intense neutrino beam than now possible anywhere.

If a factory is built, it would start off with a very high power proton beam. This beam would crash into a target, generating a huge flood of muons. And when these muons decay, there would be the neutrinos that the researchers want. But that sort of thing must already be happening in many machines where high-energy protons strike a solid target. Yes, this is happening, and neutrinos are already being produced in that fashion. But in the proposed factory, the muons would all be collected in a storage ring and even accelerated before they decay. That is where the huge gain would come from.

Conceptually, there are no unknowns in designing a neutrino factory. However, there are many technical challenges. An important one is the design of the target where the original proton beam would crash. A *tremendous* amount of heat would be generated, and this would have to be dissipated without damage to the target. A suggestion has been made that instead of a solid target, one could use a flowing liquid mercury target. Good idea, but even this requires an elaborate design.

OK, just because the concept is clear and the factory is in principle buildable does not mean it will be built. In experimental high-energy physics, ultimately it all comes down to money (!) and money is always in short supply. At today's prices the factory would cost about $ 3 billion and that is not exactly a small sum of money. Will the funding come or will the factory remain an idea on paper? Time and the taxpayer will tell!

1975–1977: Tau neutrino and b quark detected, revealing a third generation of leptons and quarks.

1979–1985: A theory that matter can enhance neutrino oscillations is developed.

1987: Super-K detects neutrinos from SN 1987A.

1989: CERN results establish that there are only three active neutrino generations.

1998: Super-K evidence for neutrino oscillations from a study of cosmic ray produced atmospheric neutrinos.

2002: SNO produces the 'smoking gun' evidence. Neutrino oscillations now established from a study of solar neutrinos.

Note added in proof: A neutrino observatory is being built at the US South Pole station. The aim is to detect extraterrestrial neutrinos with energies above 100 GeV. Antartic ice is very clear and absorbs much less light than sea water. A large number of detector modules have been lowered into the icecap to a depth of about 2 km. The Icecube Observatory as it is called, involves a collaboration of more than 150 scientists from eight countries.

12

WHAT WAS IT THAT BANGED?

The basic recipe for describing the early history of the Universe is to use physics valid at small scales of length. Even if one does that, one faces many problems. Two important ones are what are called the horizon problem and the flatness problem. This chapter describes a novel idea that gave the first clue as to how these difficulties could be circumvented in model building.

12.1 Introduction

Three major predictions follow from the so-called Standard Big Bang Model of Cosmology. They are:

- Following the Bang, the Universe started expanding, eventually making the galaxies fly away from each other.
- About 100,000 years or so after the Bang, electromagnetic radiation began to emanate from the remnants of the original cosmic soup. (Some people estimate this time to be as high as 300,000 years.) This radiation then filled the entire Universe. It has a blackbody spectrum. As the Universe expanded, its temperature began to decrease.
- The light elements, namely hydrogen, deuterium, helium and lithium formed from the cosmic soup, and the standard Big Bang Model of Cosmology even gives estimates of the amounts so formed (as already noted in an earlier chapter).

All these predictions have been essentially confirmed, making us believe that the Universe most likely originated in a Bang. Of course, we need a much more detailed picture of all these early events; this chapter will deal with an important aspect of that early history.

Now the model as given originally by Gamow and some of those who followed, relied mainly on nuclear physics. This took us back from the present time to about one second, say, after the birth of the Universe. What happened still earlier? For finding that out we have to go beyond nuclear physics. With the Standard Model of particle physics one could go comfortably up to say 10^{-12} s. To go beyond, one must appeal to GUT, and to go right up to Planck time of 10^{-43} s and possibly even beyond, one must appeal to the dream unified theory that combines GUT (which we still do not have!) with gravity.

OK, suppose one does all this. How does one know that one has got it right? Well, making up a story is not enough. One must use the model to make *predictions* that can be verified by looking out into the Cosmos *today*. In other words, any model of the *early* Universe must be able to account for what we see *today*. If it does, then the model is acceptable; otherwise we just discard it.

12.2 Two big problems

Cosmologists started off by using the SM to make predictions about the early Universe but they all ended up with two major difficulties. They are called respectively the *horizon problem* and the *flatness problem*. Let us find out what exactly they are, before proceeding further.

Astronomers find that the Universe is substantially uniform. What does one mean by this? Well, if for example we look at the microwave background radiation, we find that the radiation coming from different directions of the sky has the same intensity and also the same spectrum (which essentially means the same temperature). There are actually minute fluctuations, but we shall ignore them for the present. By and large there is uniformity, and that is what we are focussing on at present. There is also another piece of macroscopic evidence. Suppose we divide space into cubes, each side of the cube measuring say about 300 million light years. Astronomers have found that each such cube closely resembles the others in all its properties like average matter density, galaxy density, and light output. That is another reason for saying that the Universe as we see it today is rather uniform. By the way, the minute fluctuations that I chose to ignore are very important and very much needed; without them we would be unable to explain galaxy formation; more about that later, in another chapter.

OK, the Universe is substantially uniform, at least on large scales; so what's the problem? The problem is, how did it get this uniformity? If it is uniform today, it must have been so when it was a Baby Universe, but at that time the conditions did not permit any such uniformity. How and when did this uniformity come about?

You may say, "Wait a minute. I don't understand why the Universe was not in a position to achieve uniformity when it was a tiny baby." A good point and that is what we shall consider now. Let us go back to the time when the Universe was just 10^{-35} s old. Let us call this age t_B, meaning the age of the Baby Universe. Let us say it had a radius R_B. We have a good estimate of R_B; it is about 1 cm. Now let us calculate the distance that light can travel in 10^{-35} s. It is 3×10^{-25} cm, which is incredibly small compared to the diameter of the Baby Universe. You may say, "So what?" Well, that exactly is where the problem lies! You see uniformity comes from the mixing, or to put it more elegantly, when there is thermal equilibrium. By this we mean that there must be a complete exchange of information between different parts of the Universe; remember information *cannot* travel faster than light. So if light could not cross the Baby Universe when it was 10^{-35} s old, then different bits of the Baby Universe could easily have had different conditions,

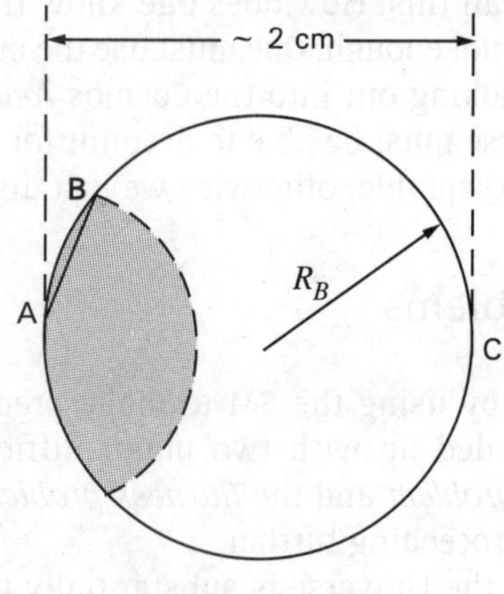

Fig. 12.1 *This figure (which, note is* **not** *to scale) shows schematically the state of affairs in the Baby Universe at time $t = 10^{-35}$ s. Due to expansion after birth, R_B is 1 cm. However, from a typical point like A on the boundary, light can travel a maximum distance of only 3×10^{-25} cm. This distance (shown by AB), called the horizon distance, is clearly very small compared to the dimensions of the Universe (at that time). Thus, around A, equilibrium is restricted to the shaded region. Hence, thermal equilibrium across the entire Universe is inconceivable. The question now is: If this be so, how come the Universe appears so very uniform at the present time? This is the so-called horizon problem.*

and that means no uniformity. So the question boils down to: "How on earth did the conditions become uniform in the Baby Universe? Without uniformity then, we would not have uniformity now. So what magic did God perform then?" What I have just described is referred to as the horizon problem. See also Fig.12.1. By the way, in this context, the word *horizon distance* at any particular time means the distance light would have travelled from the time the Universe was born to that particular instant. In Fig. 12.1, AB is an illustration of what the horizon distance means at 10^{-35} s.

We turn next to the flatness problem. As I mentioned earlier, the Universe, after having been born, has three basic options as far as its future is concerned; these are governed by the value of Ω. Astrophysicists tell us that Ω today has a value somewhere between 0.5 and 1.5. You might wonder, "This is horrible; such a wide margin and how imprecise!" We shall consider that issue later, but meanwhile, let us accept this broad band. Now this is what I would like you to note.

According to theoretical Cosmology, if the Baby Universe starts off with a value of Ω even *slightly different* from 1, then the deviation from 1 would grow rapidly with time. However, if the value of Ω at start was 1 or not different from 1 *except in the fifteenth decimal place or so,* then it would be so today also (i.e., have a value ~ 1). What it means is that though our estimates of Ω are still rough, we know enough to say that the value of Ω today is probably 1 and that that must also have been its value when the Universe was just 10^{-35} s old. How come the Universe happened to choose this special value way back? This problem of the value of Ω is referred to as the flatness problem because in Cosmology, Ω = 1 means a flat Universe. In other words, how come the Universe decided to be flat way back? How could it have been flat when it was incredibly tiny? Surely, a bubble with a radius of 1 cm cannot be described as having a flat surface. Is there some physical effect or principle that made it so?

12.3 Guth's inflationary model

OK, so we have these two problems and any application of the SM to describe the Baby Universe must be made in such a way that these two problems are taken care of. (Actually, there are other problems also but we skip them for the moment. The two problems just mentioned were considered by many as the principal bugbears and that is why I am giving them a special mention.)

In 1981 Alan Guth proposed what is called the *Inflationary Model* of the Universe that used the SM but also invoked an ingenious idea that took care of the two problems just mentioned. So what did Guth do? He said, "Listen, at 10^{-35} s the Universe did not quite have a radius R_B. The radius was in fact *much smaller,* being around 10^{-50} cm." You might say, "This is crazy; if the Baby Universe was that small then, the Universe as we see it would be tiny compared to what it actually is. Even with all the terrific expansion that is supposed to be taking place (about which we know from Hubble's work), the Universe today would not be even one mm big! Either this is a joke or there is something seriously wrong somewhere." That is where sheer brilliance comes into play!

You see, if you and I can raise such an objection, surely Guth must have seen it even before he put pen on paper to work out the theory. But he had a neat way of bypassing the problems. He said (effectively): "Let us for a moment just accept that at 10^{-35} s or so, the Universe was indeed very tiny. Next let us suppose that due to a special mechanism the Universe simply blew up from 10^{-50} cm to about 1 cm in say just about 10^{-32} s or something like that." Imagine that! An expansion by *50 orders of magnitude in just 10^{-32} s!* What is that but an explosion?! Guth called this inflation, but in practical terms this is nothing short of an explosion of cosmic magnitude; and that is why some call this inflation as the real Big Bang. By the way, there is an important difference between an explosion and inflation. In the usual explosion (like a Diwali cracker, for example), matter gets scattered, that is to say some particles are thrown afar while others do not get that far. In inflation the system gets blown up uniformly, rather like a balloon. And remember, what is expanding is the spacetime itself!

Obviously, one can't quite pull off a trick like this unless there is a sound physical basis for this, and that is what Guth provided. That is the real merit of his model; it was not just a wild idea but a well-argued model based on sound physical principles. So let us take a brief look at the basic idea underlying the inflationary model of Guth.

Inflation, said Guth, is what happens when the Higgs field of the Universe undergoes a phase transition. We have been through phase transitions: the potential well that controls the transition and all that (see Volume 1, Chapter 10, especially Fig. 10.11). For our present purpose we recall the nature of the potential well which is shown in Fig. 12.2. When the temperature T is above T_c,

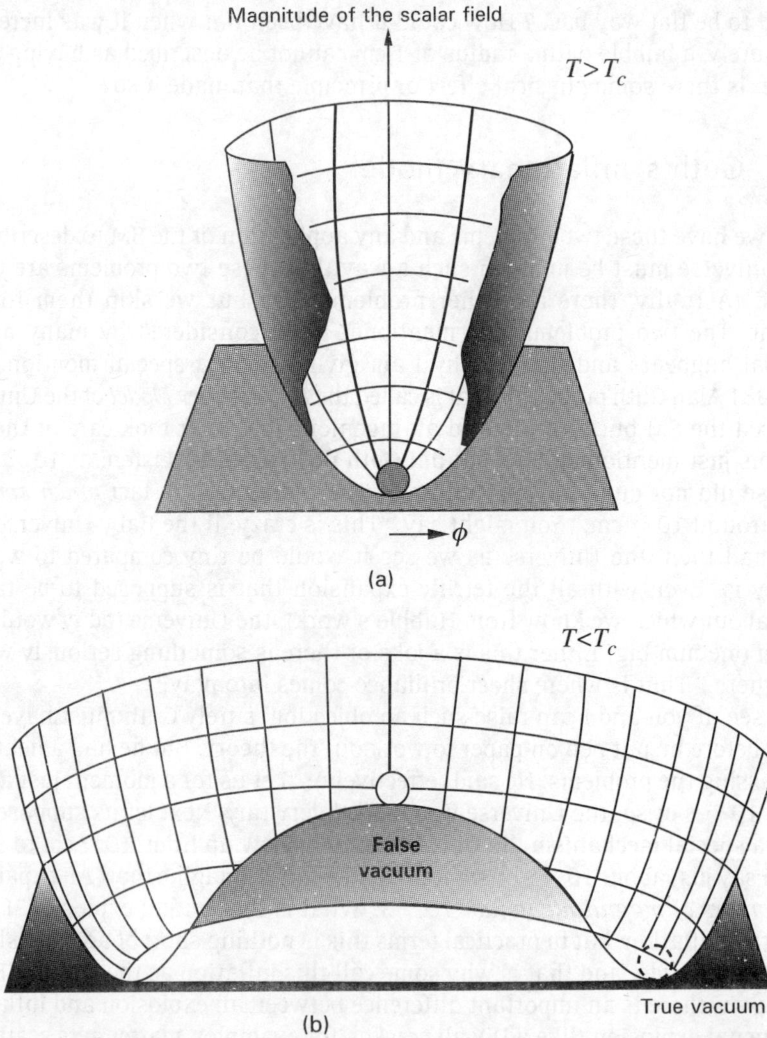

Fig. 12.2 *This shows a plot of the energy density versus the value of the Higgs field ϕ as the Universe cools from a temperature above T_c [see (a)] to a temperature below T_c [see (b)]. In (a), the minimum is at $\phi = 0$ and this minimum is the true vacuum. In (b), $\phi = 0$ represents a false vacuum while at true vacuum ϕ has a finite value. When the system gets supercooled, it initially stays at the value $\phi = 0$ where it was earlier. Later, it makes a transition to the real vacuum and that is when the energy associated with the false vacuum gets released, creating the particles in the original cosmic soup. Detailed considerations show that when the system is supercooled and is lingering around in the false vacuum state, it experiences a strong repulsive force which produces the inflation. Thus, the inflation takes place during the brief interval when the system is still hanging around the false vacuum.*

the transition temperature, the system would be in the state of zero field, which is the vacuum state. At that temperature, this state would be a perfectly stable state – this I have explained before (in Volume 1). When T falls below T_c, the shape of the potential changes from that in Fig. 12.2(a) to that in 12.2(b). The system still continues at the state of field zero. However, this is no longer the true minimum of energy, and therefore the system is said to be in a false vacuum. In physical terms, the system temperature is below T_c but it still hanging around in a state that is appropriate to a temperature T greater than T_c. What this means is that the system has supercooled. For example, when water is cooled below 0 °C, we expect it to become ice. But under controlled conditions water could be cooled as much as 20 degrees below zero and still remain water. This is supercooling. If you remember, the bubble chamber works by exploiting supercooling.

OK, the Universe has cooled down to a temperature below T_c and is in a supercooled state, i.e., it is still in a state appropriate to that for $T > T_c$. Obviously, this supercooled state cannot last for ever, and the system must, sooner or later, go over to the state appropriate for $T < T_c$. In terms of energy, it means that the system must go over from the false to the true vacuum, a process that is expected to take some time (even if it is as small as 10^{-32} s). Till that changeover occurs, the system would be in the false vacuum state. Let us examine what could happen to the Universe during the brief period when it is in the false vacuum.

Note first that when a system is truly in the vacuum state, its energy cannot be lowered. When the system is in the false vacuum state, its energy remains steady for a while till the transition occurs; that is why the word vacuum is used. However, since the state does not remain so for long, it only gives the appearance of vacuum (for a short while); hence the name false vacuum. This is the first point.

We must next note a rather peculiar property of this false vacuum which relates to its pressure. See Fig. 12.3. Imagine a chamber filled with false vacuum as in the figure. Let us say the volume is V and that the energy density is u_f. This means that the energy inside the chamber is $U = Vu_f$. Suppose now that the piston is quickly pulled outwards, increasing the volume by dV. If we did such a thing with a normal chamber filled with gas, the energy density would decrease. The false vacuum, however, cannot rapidly lower its energy density; instead, the energy density remains constant. Since the volume has now increased by dV, it is clear *the energy inside the chamber would have to increase* and its value would now be $U + dU$, where $dU = dV.u_f$. Where does this extra energy dU come from? From 'outside'. This means that the inside 'sucks' energy from the 'outside' and *this suction acts like a negative pressure* (as explained in Fig.12.3). All this is just to say that associated with the false vacuum there is a negative pressure.

Now according to general relativity, pressure produces a gravitational field. Normally we have positive pressures, and they lead to attractive gravitational fields. But when the pressure is negative (as in the case of the false vacuum), a *negative gravitational field* is produced, that is, a field which is *repulsive. In other words, during the brief period the system is in the false vacuum, the*

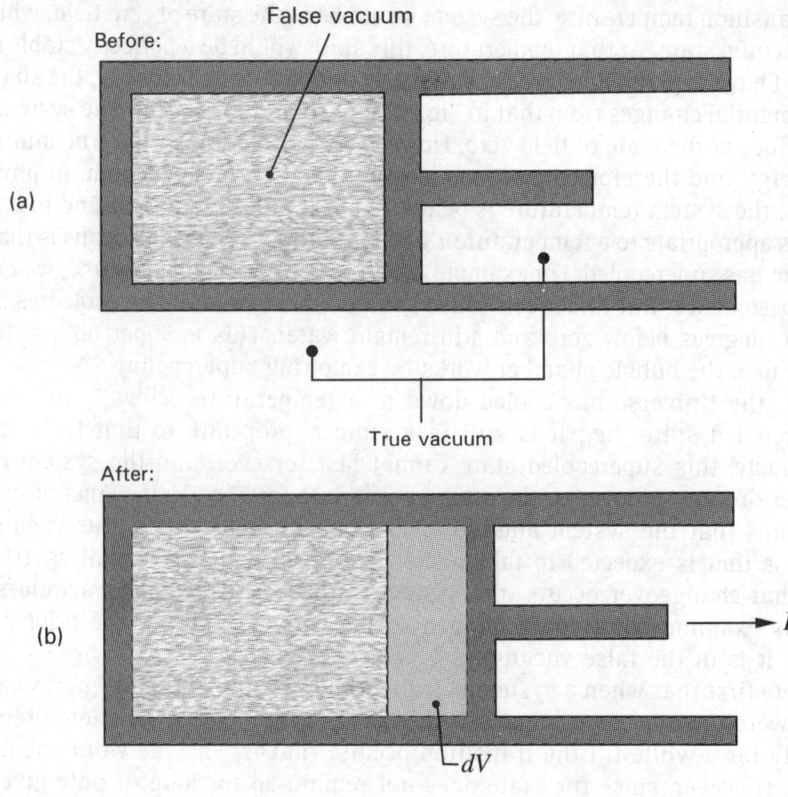

Fig. 12.3 *This figure is meant to illustrate how false vacuum generates a repulsive force leading to the rapid expansion or inflation of the Universe. Consider a chamber with a piston as in (a). Suppose the piston is quickly pulled outwards, increasing the volume by dV as in (b). The work done $dW = -pdV$. This dW must be equal to dU the change in the energy within the chamber, which is equal to $u_f dV$. This gives $p = -u_f$. This negative pressure generates in general relativity a repulsive field, and it is this repulsive field that drives the rapid expansion of the Universe.*

associated negative pressure generates a repulsive field that effectively causes the system to expand. (By contrast, attractive gravitation would always produce a contraction.)

Let us go over all this once more.

- As the Universe ages, its temperature starts falling.
- When T crosses T_c, the phase transition does not occur immediately; instead, the system briefly supercools.
- In the supercooled state the system is in the state of false vacuum.
- There is a negative pressure associated with this false vacuum.
- This negative pressure drives the expansion of the Universe.

- The expansion occurs extremely rapidly, lasting only for an incredibly short time.
- After this duration the system makes the transition from the false to the real vacuum.
- The rapid expansion now stops.
- The energy difference between the state of the false vacuum and the real vacuum gets released. This is like the release of latent heat in normal thermal physics.
- This 'latent heat' (which is actually a *tremendous* amount) is used by the Universe to generate particles (cosmic soup).

The last point is very important. *Inflation generates energy, which then creates the particles of which the Universe is made of.*

All this is fine but what about the horizon and the flatness problems? Inflation takes care of them of both. To see this, let us first consider a simple analogy. Let us say that there is a motorist who is travelling at a speed of say 60 kmph. In one hour he would have travelled 60 km. Suppose now that the road itself expands rapidly; in that case, along with the road, the driver also would be pushed forward. The net result is that at the end of one hour, he would be at a distance greater than 60 km from the starting point – see Fig.12.4. That is the feature of the inflation model also. As Fig.12.5 shows, just before inflation the Universe was very small and the horizon distance at that instant much smaller than the size of the Universe.

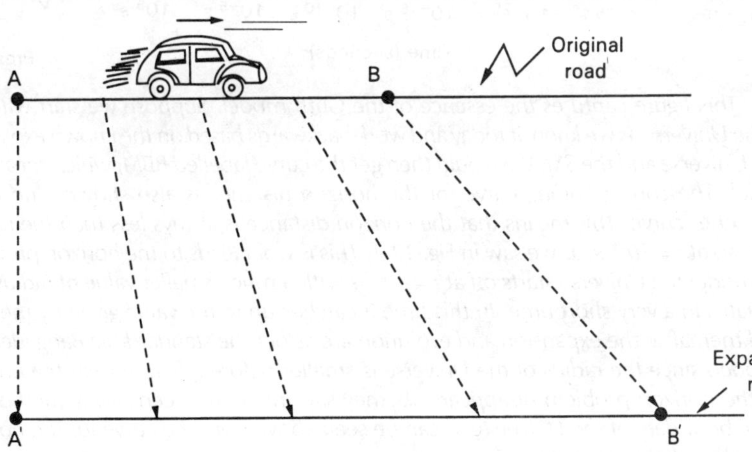

Fig. 12.4 *The purpose of this figure is to illustrate how the horizon distance expands consequent to inflation. At the top we have a road on which is travelling a car that has started from A and is moving with a uniform speed of 60 kmph. At the end of one hour the car is expected at the spot marked B which is 60 km away from A. Suppose however, the road itself expands while the car is moving. We would then have a situation like that shown at the bottom. Here, the spot B of the top figure moves over to B′ which is more than 60 km from A, the starting point. The net result is that thanks to the expansion of the road, the driver would have, at the end of one hour, effectively travelled a distance greater than 60 km though he has been driving only at a speed of 60 kmph. In Guth's inflation model, this is what happens to the horizon distance since light is like the moving car.*

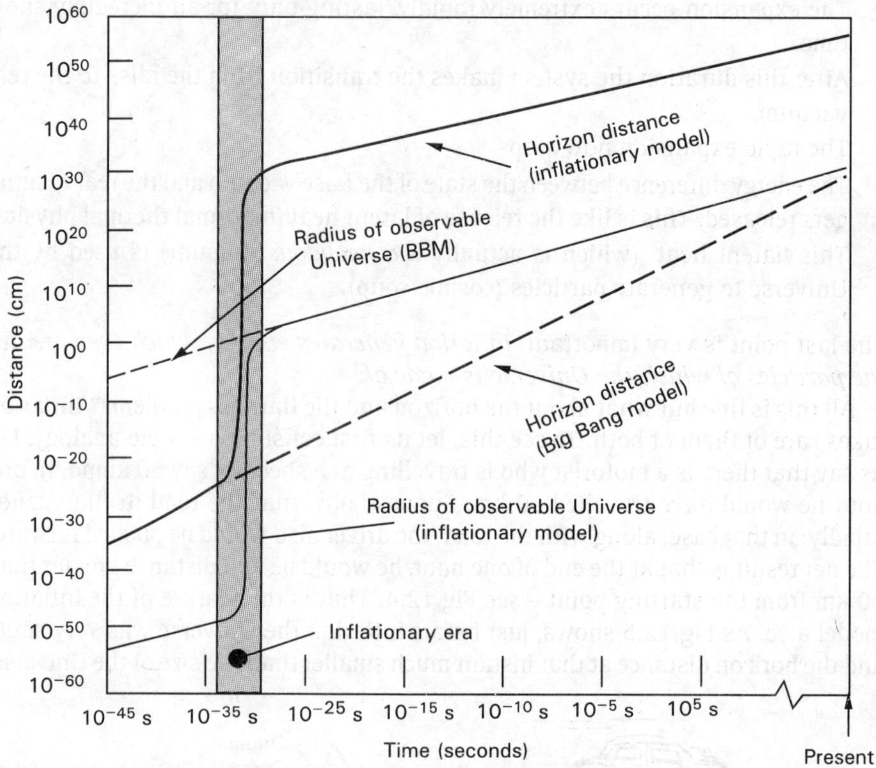

Fig. 12.5 *This figure captures the essence of the Guth model. Suppose we start with the radius of the Universe as we know it today and work backwards based on the known expansion rate of the Universe and the SM. We would then get the curve labelled BBM, which means Big Bang Model. The corresponding curve for the horizon distance is also shown, and it lies below the other curve. This means that the horizon distance is always less than the radius, particularly so at $t = 10^{-35}$ s, as we saw in Fig. 12.1. This is what leads to the horizon problem. In Guth's model the Universe starts off at $t = 10^{-35}$ s with a much smaller value of radius, but rapidly inflates in a very short time. In this time it catches up to the value given by the BBM curve, and thereafter the expansion and evolution are as per the standard Big Bang ideas. In Guth's model, since the radius of the Universe is* smaller *before inflation than the horizon distance, the horizon problem disappears. As they say, inflation sweeps away the horizon beyond the boundary of the Universe, as can be seen above from the curve for the horizon distance in the inflationary model.*

Hence, different parts of the Universe could communicate with each other and thermal equilibrium was possible. Inflation then took place, suddenly boosting the size of the Universe to the size of the standard Big Bang Model. However, there is a difference and that is with respect to the horizon distance. As in the example of the driver on the road that expands, the horizon distance also expands (light is like the driver on the road). As Fig. 12.5 shows, in practically an instant the horizon distance becomes larger than the size of the Universe and remains so thereafter. So there is never any horizon problem! Thus, as far back as 10^{-35} s or

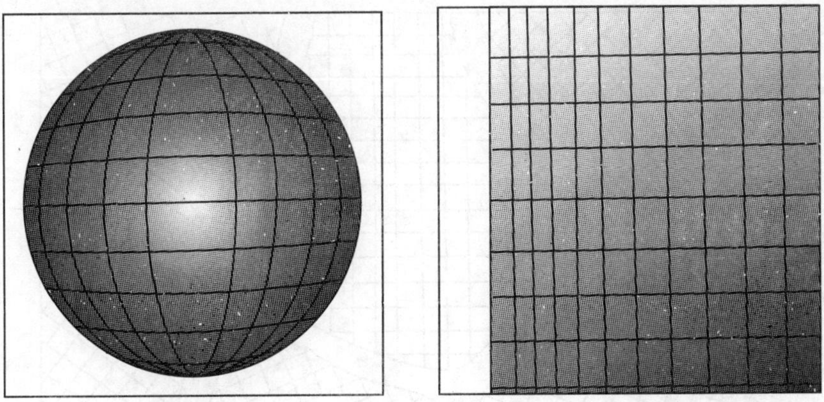

Fig. 12.6 *This figure shows schematically the Baby Universe before and at the end of inflation. Actually, it is impossible to show the Universe at the end of inflation on the same scale as that for the initial state since inflation blows up the Universe by 50 orders of magnitude. Even so, this figure is enough to convey that during inflation, the Universe expands so much that it begins to look substantially flat in local regions. In more technical terms, inflation has the virtue of driving Ω to a value close to 1, no matter what its starting value. This is what fixes the flatness problem of the standard Big Bang models.*

so immediately after the inflation, different parts of the Universe could have related to each other to establish equilibrium. In other words, inflation solves the horizon problem.

What about the flatness problem? The equations of the inflationary model show that thanks to phenomenal expansion of the Universe, Ω is rapidly driven by inflation to a value close to 1, *no matter what its initial value.* This happens during the incredibly short time taken for inflation. In other words, the Universe becomes quite flat immediately after inflation compared to what it was before (see also Fig. 12.6).

12.4 Some problems with Guth's model

Now one of the problems of the Standard Big Bang Model of the birth of the Universe is that it leaves the Universe with a lot of 'defects' that is to say, monopoles, domain walls, etc. Defects arise in almost all the phase transitions familiar to us in solid state physics. For example, when a liquid becomes a solid, though the solid is crystalline, we do not get a single crystal. We get instead a polycrystal that has defects called grain boundaries – see Fig. 12.7. The same sort of thing happens when one tries to apply the phase transition idea to the birth of the Universe in conjunction with Big Bang ideas. The original Guth model no doubt solved the horizon and the flatness problem, which showed that inflation was certainly a good idea. But it still had problems with defects: there were, for example, too many left over monopoles, and this is a problem because we hardly see any. There were a few other problems too, for example, the resulting Universe

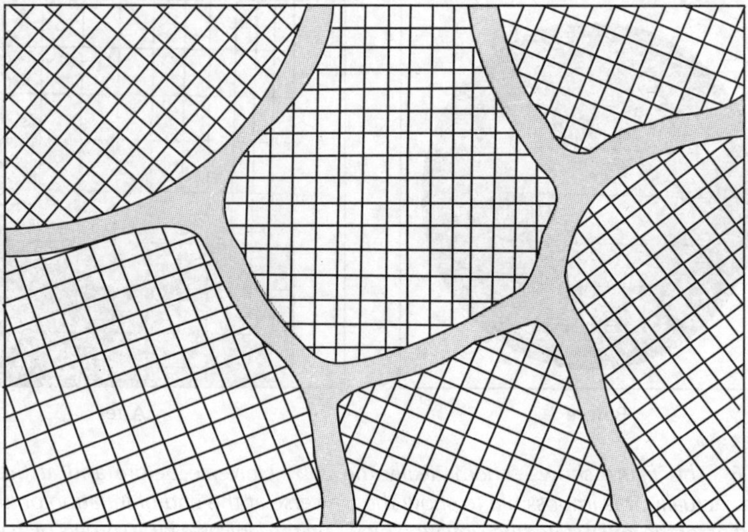

Fig. 12.7 *We all know that when a liquid is cooled below its freezing point, it becomes a solid. We also know that many solids are crystalline in nature. However, if for example we cool a mass of molten iron, we will not get a single crystal of iron. Instead we would get a polycrystal as above. This polycrystal contains a large number of single crystal grains but oriented randomly with respect to each other. The boundary between two grains is called a grain boundary. Since the atoms in the grain boundary are shared by two small crystalites with different orientations, there is a bit of disorder there. Consequently, a grain boundary is a crystal defect.*

was *too inhomogeneous*, and that contradicts the fact that the observed Universe is quite homogenous. Nevertheless, inflation did seem a good idea and was therefore widely welcomed.

Since the time of the original Guth model, many attempts have been made to go beyond and in the process also take care of some of the problems present in the Guth model. Perhaps the most daring step in this direction has been taken by Alfred Linde, then in Russia and now in America. Before we leave the original inflation model, take a good look at Fig. 12.8 which gives an overview of the old inflation model.

12.5 Linde's chaotic inflation model

We now turn to Linde, who offers us what he calls the **chaotic model of inflation**. This model is rather complicated in its details and I obviously cannot do full justice to it. However, we shall try to understand the essential features and for this purpose, we need first to remind ourselves of the basic features of the Guth-type inflation models. They are as follows.

- There is only one Universe.

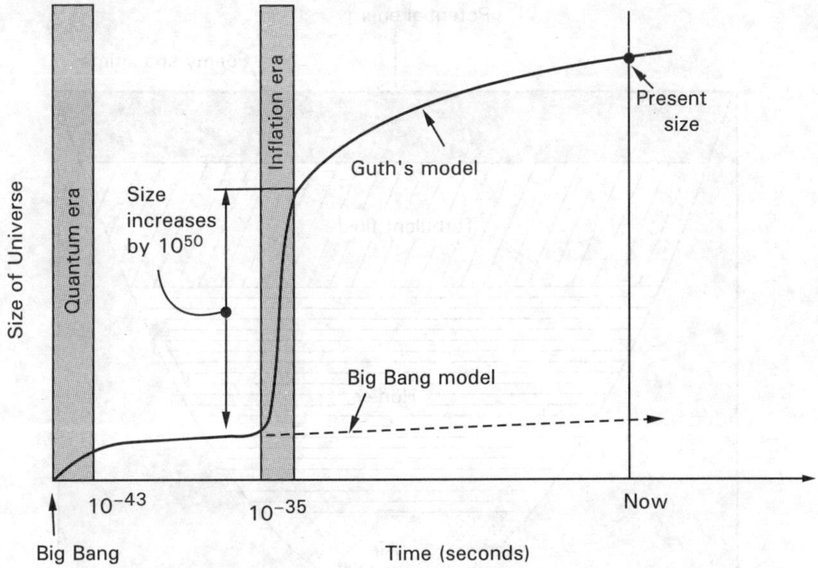

Fig. 12.8 *This figure, though highly schematic, compares the traditional Big Bang picture with that offered by the inflation model. Compare it with Fig. 12.5.*

- Initially, the energy density of the scalar field is zero.
- There is then a sudden supercooling.
- The Universe which is like a bubble, then suddenly inflates—this is what may be interpreted as the Big Bang.
- The inflation is over real quick, in about 10^{-32} s or even less.
- The stored energy is then released—this is like the release of latent heat. This energy is what manifests as particles. With this the phase transition is complete.
- The inflation over, the Universe now goes through the standard expansion of the Hubble type.

Linde breaks away from this picture substantially.

- He does not invoke a phase transition.
- He does not demand that the scalar field density initially have a value of zero.
- He gives a prominent place to *quantum fluctuations* as he feels they must play a dominant role in the early stages.

There are a number of other details but they are not of immediate concern to us. OK, at the end of it all, what is it that we get? What's the picture like? Very different, to say the least. Instead of the Universe looking like one expanding ball of hot gas, we have many inflating balls, some of which give rise to new balls, etc; and we live in one of those balls! In other words, we do not merely have a Universe but an entire *Cosmos*, made up of many balls/Universes. As Linde calls it, it is a self-reproducing Cosmos.

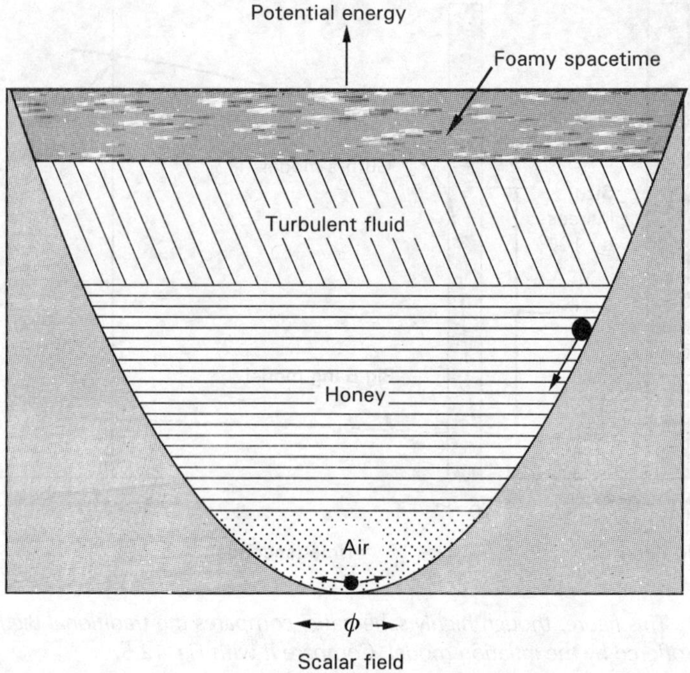

Fig. 12.9 *The purpose of this figure is to highlight the fact that in Linde's model there are many possible scenarios for inflation. Basically, expansion occurs when field energy is lowered and gets released. This lowering is visualised as a ball rolling down the sides of a parabolic bowl. However, this bowl is not a simple one, being filled with various fluids as described in the text. This introduces many variations of the scenario possible, depending on where the ball is placed for rolling down. All these different possibilities together lead to the picture of a non-uniform balloon expanding, which is illustrated in a later figure.*

The physics of the different bubbles/Universes may be different but that need not bother us. In any case, we cannot communicate with them! All we have to accept is that Destiny has placed us in one particular cosmic bubble that forms a part of the whole of Cosmos, and in this special bubble we exist! Those are roughly the headlines, if I may say so. Let us now take a slightly closer look.

We begin with Fig. 12.9 which illustrates how Linde's starting point itself departs significantly from that Guth. As usual we have a potential, and the state of the system is depicted with a ball rolling down. Note that the ball starts off from a position where ϕ is not zero. Linde just discusses the scenario associated with the ball coming down and the variations possible, that is all; no talk of phase transition anywhere.

OK, what next? Imagine the potential to be a bowl. Imagine now that the bowl is filled with several layers of fluids—mind you this may tax your imagination a bit, but if you are good dreamer you should have no problem! I want you think that at the bottom there is air, and on top of it a thick fluid like honey (don't ask me how honey can float on air; that is the price you have to pay if you want analogies!). And above the honey is a turbulent fluid! Enough to drive you crazy?

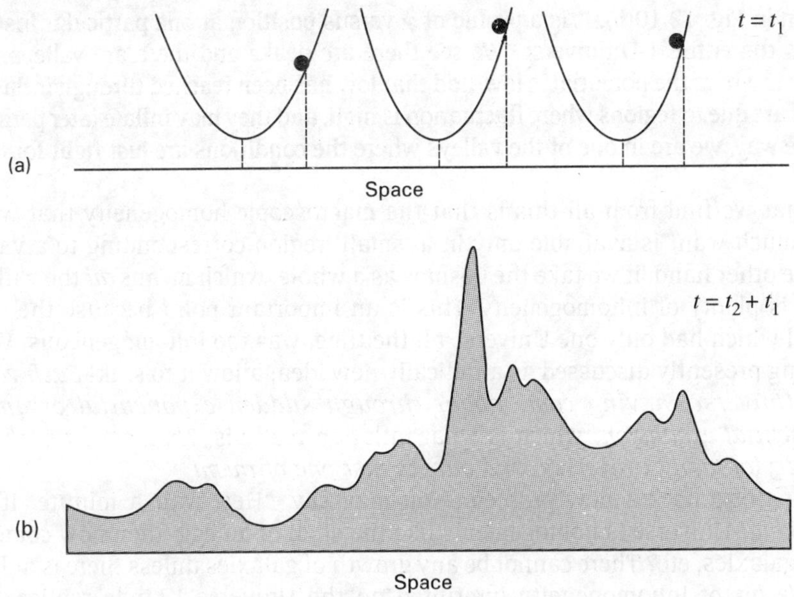

Fig. 12.10 *Let us start with (a). This shows 1-D space at a very 'early time'. Different regions have different values of the scalar potential, as indicated by the positions of the different balls in the various bowls. Evolution is now allowed to take place and we take stock a little later. The variation of the scalar field across this 1-D space (at a particular instant of time) is sketched in (b). Essentially, the curve is a locus of the ball position at various space points at time $t_2 > t_1$. In this graph, we see peaks and valleys. Valleys represent minima, and these have been arrived at by 'the ball rolling down'. They thus represent little Baby Universes which have just undergone inflation, each in its own way.*

All this would help. Let us now imagine placing a ball in the bowl, on its side. We can introduce the ball where there is air, or where there is honey or where there is the turbulent fluid. If the ball is in the region where there is air, it would quickly roll down to the bottom, rattle there and eventually settle down. If the ball is placed where there is honey, the viscosity would initially inhibit the roll down and the ball would hang around there for a while moving quite slowly; eventually it would cut free, come down to the region with air, roll down rapidly to the bottom, and then rattle there for a while before settling down at the bottom. If the ball is in the region where there is turbulence, it may even be kicked up. Above the turbulence region the energy is so high that one is in the Planck domain of spongy spacetime. What we learn from all this is that different scenarios are possible, depending upon where the ball is started off from. Keep all this in mind.

Let us now try to understand what relevance all this has for the evolution of the Universe. Consider Fig. 12.10. This shows a one-dimensional Cosmos. There are quantum fluctuations everywhere, and as a result, the state of different regions may be represented with balls at different positions on the side of the bowl. Some balls may be in the region with air, some may be in the region with honey, and some may be in the turbulent region. OK, let us allow time to pass, and ask what the status of the potential is now at different regions of the Cosmos. This is shown in Fig. 12.10(b). It is a profile of ϕ versus position at one particular instant, across the entire 1-D universe. We see there are peaks and there are valleys. The valley is where the potential is low, and that low has been reached through inflation. Peaks are due to regions where fluctuation is high, and they may inflate later perhaps. By the way, we are in one of the valleys where the conditions are just right for us to exist!

What we find from all this is that the macroscopic homogeneity that we so very much want is available only in a 'small' region corresponding to a valley. On the other hand, if we take the Cosmos as a whole, which means *all* the valleys, there is plenty of inhomogeneity. This is an important point because the Guth model which had only one Universe all the time, was too inhomogeneous. What is being presently discussed is a radically new idea; allow it to sink! *Each valley is a Universe, having come about through sudden exponential or quasi-exponential expansion,* which is what inflation really is. *The Cosmos is thus a tapestry of many Universes, and ours is just one of them!*

But this raises a new problem. You may say, "Hey, wait a minute. If the *individual* Universe is homogeneous like the shell of an egg, then how come we have galaxies, etc? There cannot be any growth of galaxies unless there is at least a *little bit* of inhomogeneity imprinted on the Universe." Linde replies, "No problem! You see, my model calls for quantum fluctuations. So while the ball is rolling down through the honey and later through the air, there are little bumps it

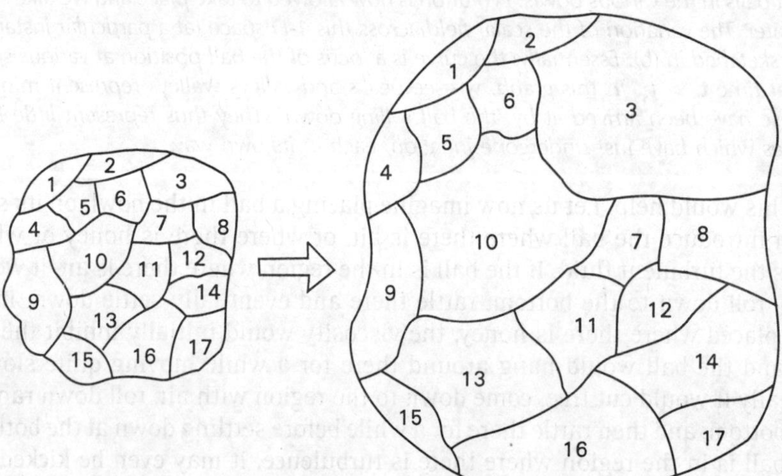

Fig. 12.11 *This figure seeks to highlight the role of quantum fluctuations in producing mild inhomogeneities. On the left is a small patch that depicts the state of the Universe at some point of time. Due to quantum fluctuations, there could be mild inhomogeneity when the Universe inflates and subsequently expands, with different pieces of the patch on the left undergoing different amounts of expansion. That is to say, different regions of the Universe might inflate slightly differently. The imprint of this inhomogeneity could even be the seed for galaxy formation. The point is that while inflation no doubt produces overall homogeneity, there are still some small quantum fluctuations in temperature and density and these are crucial for the seeding of galaxies.*

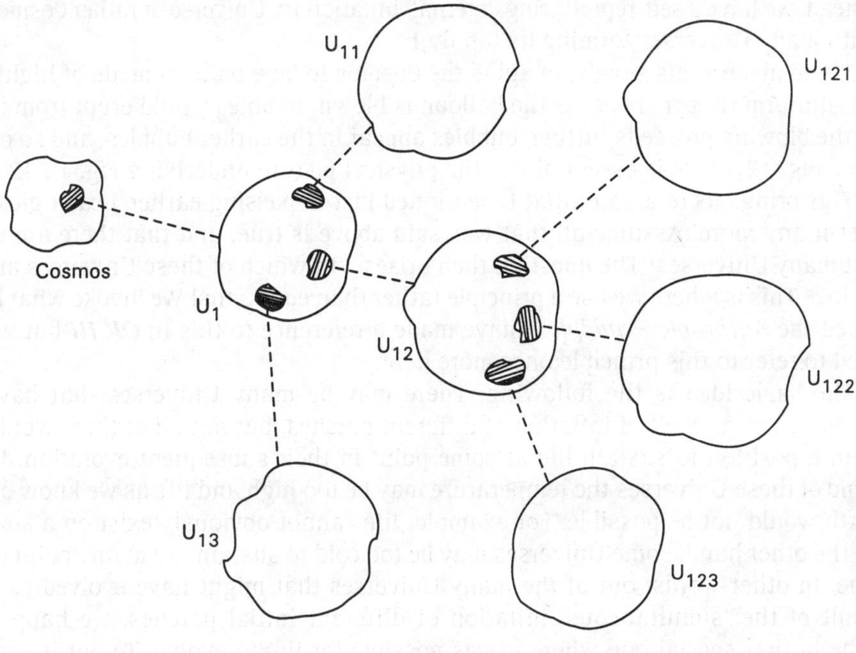

Fig. 12.12 *Shown here schematically is how quantum fluctuations lead quite naturally to an eternally inflating Cosmos. Here, time generally flows from left to right. Let us start from one patch of the Cosmos that is shown shaded. This develops into the Universe U_1, say. From this emerge three Universes, U_{11}, U_{12} and U_{13}. And from the one in the middle of this trio emerge three more, i.e., U_{121}, U_{122} and U_{123}. It is like a couple having children, grand children, great-grand children and so forth. The process keeps on going endlessly! The Cosmos is the totality of all the Universes.*

faces due to quantum fluctuations. These are enough to imprint that wee bit of inhomogeneity that you want." See also Fig. 12.11.

Fine, but what about the business of self-reproduction and all that, mentioned earlier? How does that come about? Let us consider one Universe in the tapestry called the Cosmos. Even in this Universe, during its evolution, especially in the early stages, there would be quantum fluctuations. In fact it is these that later promote the formation of galaxies. These inhomogeneities are small stuff; let us look for something much bigger. Suppose that in one spot in the Universe there is a violent fluctuation raising the potential locally. (Could be this spot is a black hole that has shrunk to a very small size, say to Planck dimensions.) What goes up must come down, and so when this potential (that has locally gone up) tries to come down, that region suffers inflation. In other words, a 'bubble' is born in that region where there was a violent quantum fluctuation, and this later develops into a Universe of its own.

This is a very interesting idea because it essentially represents a *new* Universe emerging from a region of the 'old' Universe. I am not sure if you have grasped all that I am trying to say. Even if you haven't, the gist of what I am saying should be slightly clearer from Fig. 12.12. You will realise that at this stage we have moved very far away from the idea of *one* Big Bang and *one* Universe!

Instead, we have a self-reproducing, 'eternal' inflationary Universe or rather Cosmos (with many Universes forming its family)!

Try to absorb this slowly. Imagine the Cosmos to be a balloon made of highly non-uniform rubber sheet. As the balloon is blown, bubbles would erupt from it. As the blowing proceeds further, bubbles appear in the earlier bubbles, and so on – see Fig. 12.13. This essentially is the physical picture underlying Fig. 12.12.

This brings us to a point that I mentioned just in passing earlier. I can't gloss over it any more. Assume all that was said above is true, and that there are in fact many Universes! The question then arises: "In which of these Universes are we in?" This is where we use a principle rather than equations! We invoke what is called the *Anthropic Principle*. I have made a reference to this in ***QR III*** but we need to refer to this principle once more here.

The basic idea is the following. There may be many Universes that have developed as a result of inflation of different patches, but not all of them would be in a position to sustain life at some point in their subsequent evolution. In some of these Universes the temperature may be too high and life as we know on Earth would not be possible. For example, life cannot obviously exist on a star. On the other hand, some Universes may be too cold to sustain life at any point of time. In other words, out of the many Universes that might have evolved as a result of the 'simultaneous' inflation of different initial patches, we happen to be in that special one where it was possible for life to evolve. To put it very roughly and rather crudely perhaps, our Universe exists because we are there to see it! That may be, but is there any scientific criterion we can apply to describe the special nature of our Universe? Yes, it turns out that our Universe has a value of Ω close to one; other Universes might not have this value. We are in that Universe or bubble, where we can be, that is to say where conditions permit the evolution of life. In fact, physics itself could be very different in the different Universes, and we simply have *no idea* of what these other Universes are. Nor can we communicate with them. Looks like having the value 1 and our own existence are closely connected, isn't it?!

In a nutshell, purely from the physics point of view, all we can say with certainty is about the physics of our own Universe. But while we cannot say much about the other Universes, this much we can say within the framework of the Linde model – our Universe is not the only one in existence. When we talk of the Big Bang, age of the Universe, etc., we must remember that all these refer to our own 'bubble' only! The 'parent' Universe, if I can refer to the entire 'family' of bubbles that way, is perhaps ageless! Why at all did God make it this way? No one knows, but we sure are left to wonder at the marvellous mysteries of the Supreme Creator!

It thus appears that perhaps Cosmos is perpetuating itself, rather like the living species of our world. I am sorry I am not in a position to give all the details of the above story and others like it—there are so many models and so many scenarios and we really do not know at present which of these models is closest to the truth. So my ignorance does not really matter! All I want you to appreciate is that extraordinary ideas are floating around, and the story of the Universe may be more subtle and complex than what we thought it was! You might wonder: Is this science or science fiction?! Even as you are wondering, you might have a specific

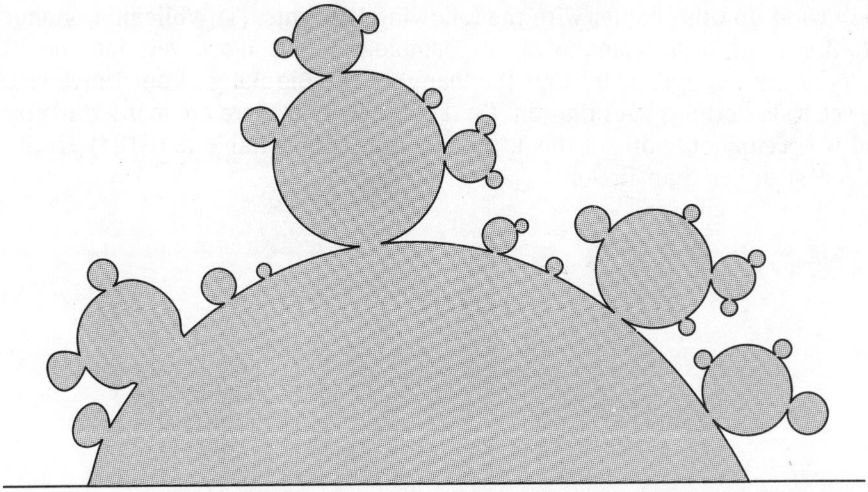

Fig. 12.13 *This figure illustrates the same concept as in the previous figure but the graphics is different. What we have here is a balloon, which, unlike the normal balloons children play with, is made of rubber that is highly non-uniform. When this balloon is blown, we will then get bubbles riding on bubbles which themselves ride on others. This is the essence of Linde's chaotic inflation model illustrated in Fig. 12.12. Shown above is a portion of the cross section of this 'bubble-decorated' balloon.*

question: "Listen, how did it all begin? Was there a primordial Big Bang out of which what you call the Cosmos was born?" Linde gives the following reply: "The situation at the very beginning is very uncertain. There is a chance that all parts of the Cosmos were created simultaneously in an initial Big Bang singularity. The necessity of this assumption, however, is no longer obvious." So it seems that, as with chaotic inflation, the question of the 'original beginning', if any, may be left to philosophers! By the way, this issue of singularity at birth is very important, and will in fact form the subject of the next chapter.

12.6 Summary and some further thoughts

- The standard Big Bang Model has many problems like the horizon problem and the flatness problem.
- Guth's inflation model gave the first clue as to how some of these problems could be circumvented.
- But Guth's model had problems of its own; this led Linde to propose the chaotic inflation model.
- In Linde's model, there is a Cosmos made up of many Universes. The one we are in happens to be the one with just the right conditions for us to live!
- Further, in Linde's model, there is an endless process of Universes giving birth to other Universes and so on, rather like the living species.

I will wind up this chapter with the following thoughts. (1) While in economics inflation is an unpleasant word, in Cosmology it is a most welcome one! (2) There is general agreement that the 'bang' of the Big Bang of our Universe as we see it, is nothing but inflation. (3) It could be that there are many Universes and we occupy only one of them, the one that is hospitable to us! (4) Truth is indeed stranger than fiction!

13

DID IT REALLY START WITH AN INFINITY?

This chapter deals with a knotty question: Did the Universe start with an infinity or did it not? It turns out that classically speaking it did, which is an embarrassment. However, if quantum considerations are brought in, then the complications of the singularity attending the birth of the Universe can be bypassed.

13.1 Introduction

In the previous chapter we focussed on a model that helps in sorting out some knotty problems like the horizon problem, the flatness problem, etc. You may ask: "OK, at some point very early in its history, the Universe underwent a terrific inflation. Fine, agreed. But what happened before that? How did it all really start? What really was the situation at $t = 0$?"

This is a tricky question that has received some attention from time to time. No doubt there were many who were not so much bothered about the $t = 0$ situation as much as what happened later. But there were others, perhaps only a few, who were not prepared to sweep the problem of initial conditions under the rug. Well, what was it that bothered these people? The issue was this.

They said (in effect): "Listen, it is all nice to talk of the Big Bang, the models of Friedmann, and so on. But if you pause for a moment and ask about the condition of the Universe at the moment of birth, at *exactly* $t = 0$ if you like, do you know what you get? You get a state of infinity, that is, the Universe started with infinite density." You may ask, "So what?", but you see theorists abhor infinity where physics is concerned. Thus, for these 'philosopher-types', it was unthinkable for the Universe to start from a singularity; it was unthinkable because no one knows how precisely to handle a singularity. Many felt that perhaps Nature had circumvented the initial singularity in a very clever way; if so, one must find out how it was done.

It is important to appreciate the spirit of these people. They were not aimless philosophers but the conscience-keepers of physics one might say. They just could not accept an infinity or a singularity as the starting point of the Universe; it went against their grain. These people just could not sleep unless the singularity at the time of birth which most theories seemed to point at and almost everyone was ready to swallow, was got rid off. This chapter is about where we stand today in this matter.

13.2 A bit of black hole physics

It all started with black holes. I have already introduced you to black holes in ***Chandra***. Briefly, when a massive star with a mass far exceeding the Chandrasekhar limit runs out of fuel, it starts experiencing gravitational collapse under the influence of its own mass. In such a star light is trapped and cannot escape, and that is why it is called a black hole. All this I have explained elsewhere. What we are now interested in is the ultimate fate of this black hole. It was held by many that the gravitational collapse of the black hole would go on endlessly, till the defunct star shrinks literally to a point. If all the huge mass of the star that once was is now concentrated at a geometrical point, then it means that the ultimate fate of the star is a singularity.

Singularity—that is a bad word. How can we have a singularity in physics? Does the black hole *really* end up in singularity? In 1965, Roger Penrose (you have met him in ***Many Phases***) did a careful analysis and came to the conclusion that if general relativity (GR) is correct—and we believe it is at the classical level—then the black hole *must* end in a singularity. Stephen Hawking read about all this. He asked himself a simple question that effectively went like this: "OK, matter can collapse due to GR into a singularity. Let us accept that for a moment. Suppose I reverse the direction of time in Penrose's theorem so that the collapse becomes an expansion. Can an expansion start off from a singularity? In particular, can the Universe start from a singularity?" Hawking examined the issue thoroughly and was forced to conclude that if GR is true, the Universe has no option but to start in a singularity. In fact, Hawking and Penrose later collaborated and produced what was considered the definitive paper on the subject. It simply said, "Sorry, but according to GR the Universe *must* have started off with a singularity." But such was the historical prejudice against singularity as the origin of the Universe that the matter did not rest there. And the man who showed how one can get past the singularity associated with the origin of the Universe was Hawking himself!

For this part of the story we have to start once again with the black hole. In the early seventies of the twentieth century, an interesting discovery was made about black holes. To appreciate this we must first remind ourselves about the *event horizon* associated with a black hole, a concept that is illustrated in Fig. 13.1. Basically, the event horizon is a unique surface associated with a black hole whose area depends on the mass of the original collapsing star. The discovery that was made in the early seventies concerns the changes that occur to the event horizon when additional matter falls into the black hole. In particular, it was found that if two black holes collide and merge to form a single black hole, the area of the event horizon around the resulting black hole is greater than the sum of the areas of the event horizons around the original black holes – see Fig. 13.2. This property of black holes rang a bell, for it seemed very much like the property of entropy that one encounters in thermodynamics. There, if two systems (say two gases) with entropies S_1 and S_2 say, mix together, the entropy S of the resulting system is greater than $(S_1 + S_2)$. Something similar was found by Bekenstein in 1972 in relation to the areas of event horizons.

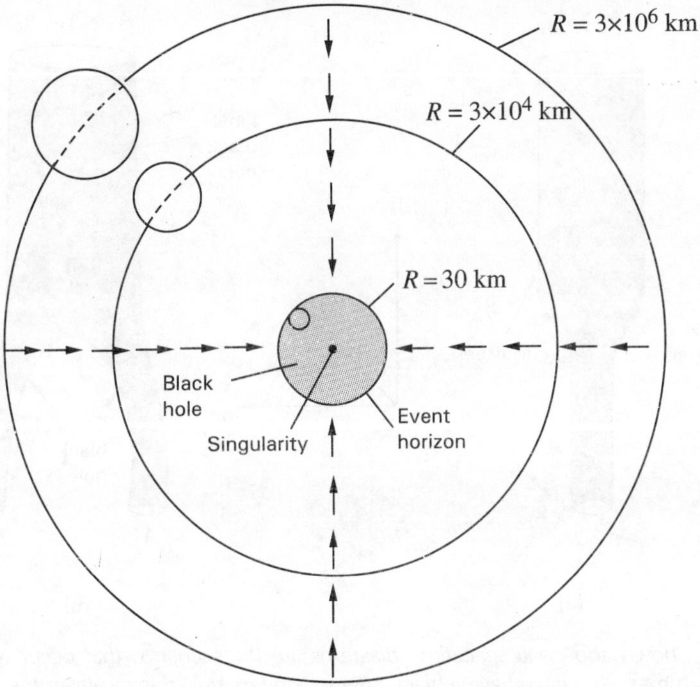

Fig. 13.1 *The aim of this figure is to explain the meaning of event horizon. Suppose a star with a mass about ten times the mass of the Sun and a radius of say about 3 million km undergoes gravitational collapse. Shown above are three stages during the collapse. The small circles represent the size of a wavefront of light emitted just before the collapsing star reached the size indicated. As the collapse proceeds, more and more of the wavefront is trapped inside the mass of the star, indicative of the increasing influence of gravity. When the star reaches the size of the event horizon (shown shaded), light fails to escape outside.*

Now if you start talking about entropy of a black hole, then you must also associate a temperature with the black hole; mathematically, this could be done. However, there was a problem. In thermodynamics, a hot body radiates energy. So, if the black hole had a temperature associated with it, it had to radiate. But the black hole was supposed to swallow photons and not emit any. So how on earth could the hot black hole radiate energy, as thermodynamics required it to? This was something of a paradox, and the man who resolved it was Hawking.

Hawking said (effectively), "Well, it is time to stop thinking of the black hole purely as a classical object and start applying a bit of quantum mechanics to it." Hawking did precisely this, and lo and behold, he found the black hole could radiate energy! This may sound unbelievable but quantum physics does provide a nice loophole. Thanks to this quantum mechanical loophole, while photons cannot get out of the black hole, the latter can still radiate energy! How did Hawking pull this trick? Using the idea that the vacuum can generate particles!

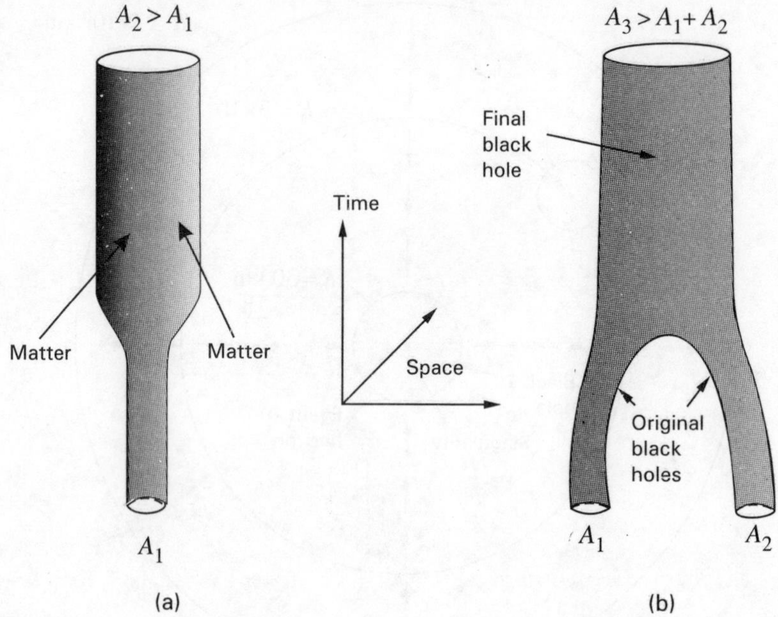

Fig. 13.2 *Shown above via spacetime diagrams are the scenarios that occur when black holes absorb mass. In (a) we have a black hole sucking matter; this results in the increase of the area of the event horizon. In (b) we have two black holes colliding and merging to become a bigger black hole. In this case the combined area is greater than the sum of the areas the colliding black holes before the merger. It is this property of the area that suggested a link with thermodynamics, which Bekenstein and Hawking gave proper shape to.*

Let me go over this part of the story slowly. We start with Fig.13.3 which shows how the vacuum can be busy as a bee. Cut now to Fig. 13.4(a), which shows the happenings in the vacuum near the edge of the black hole; we see that in the vacuum neighbourhood, virtual creation of particles–anti-particles goes on all the time (as indeed it does elsewhere as well). In some cases the anti-particle member of the pair falls into the black hole and is lost for ever while the other partner escapes to infinity.

OK, so what? Well, this is where we use a trick taught to us by Feynman and redraw some of the events of Fig. 13.4(a). This is done in Fig. 13.4(b). What do we find? Effectively, we have everything described in terms of the particle alone. The particle first travels back in time, out of the black hole into the neighbourhood. When the particle reaches the point where the pair was materialised, it is scattered by the gravitational field whereupon it now begins to travel forward in time, away from the black hole, into outer space. This is what observers outside the black hole see.

Let us pause and absorb. QM has seemingly allowed a particle to escape from the black hole! One way of looking at this entire process is to say: "Pair production took place *outside* the black hole. One member of this pair fell into the black hole and the other member escaped and is seen by an observer far away as coming towards him". However, thanks to QM and Feynman, we can reinterpret the entire

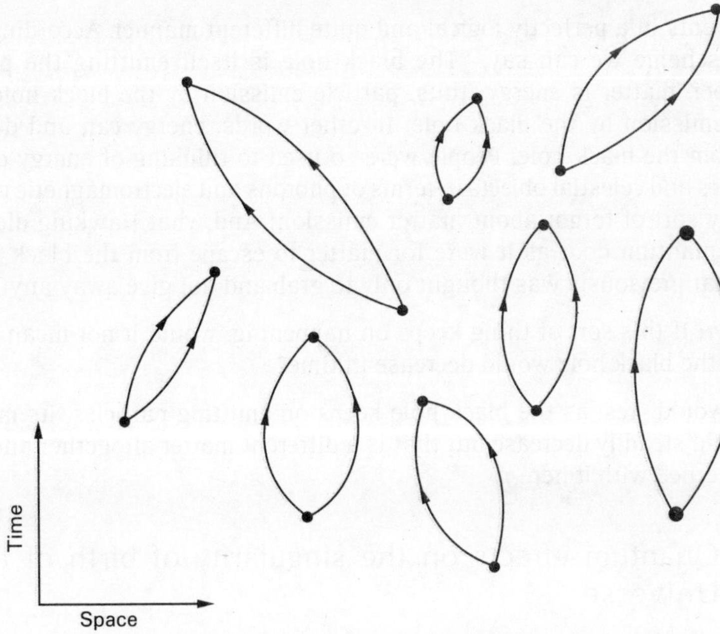

Fig. 13.3 *This figure is merely meant to recall something that you know, namely that in the so-called quantum vauum, there is virtual pair-production going on all the time.*

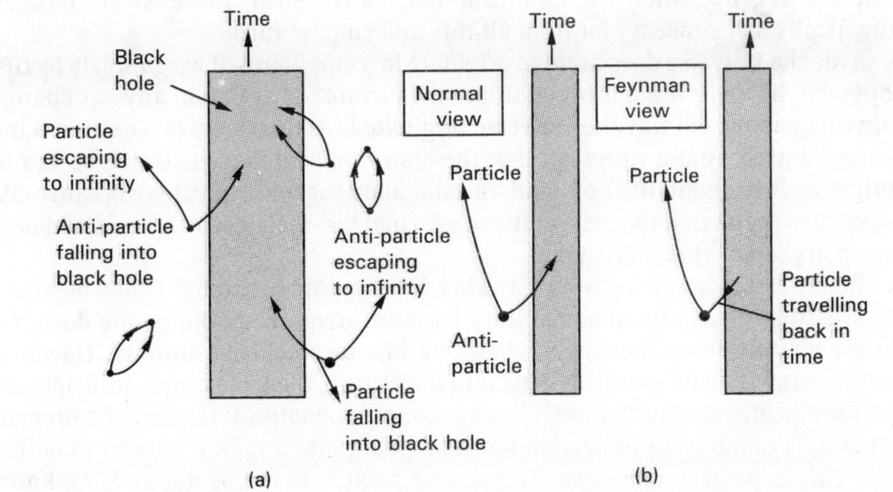

Fig. 13.4 *This figure, like the previous one, is also a spacetime diagram. In (a), we have an illustration of the pair-production in the vicinity of the event horizon of a black hole. Notice that in some cases, one of the particles belonging to a pair enters the black hole to disappear there for ever. In (b), this story is retold by appealing to Feynman's trick relating to travelling back in time. As can be seen, in the Feynman picture we essentially have a particle that escapes from the black hole to the outside. It is this that is interpreted as the emission of energy by the black hole.*

set of events in a perfectly logical and quite different manner. According to this second scheme we can say, "The black hole is itself emitting the particle". Remember, matter is energy; thus, particle emission by the black hole means energy emission by the black hole. In other words, energy can and does flow away from the black hole. People were so used to thinking of energy emission from stars and celestial objects in terms of photons and electromagnetic radiation that they sort of forgot about matter emission! And what Hawking did was to open a quantum door as it were for matter to escape from the black hole, an object that previously was thought only to grab and not give away anything.

Question: If this sort of thing keeps on happening, would it not mean that the mass of the black hole would decrease in time?

Sure it would. Yes, as the black hole keeps on emitting particles, its mass and size would steadily decrease but that is a different matter altogether and we are not concerned with it here.

13.3 Quantum effects on the singularity of birth of the Universe

Now why did I bring in all this? What if the black hole radiates? What has that got to do with the problem of the Universe originating in a singularity? Simply this. When black holes were looked at purely from a classical point of view, we got one answer, but when the quantum spice was added, the answer changed drastically. Let's take a hint from all this and enquire further.

True the Universe does seem to originate in a singularity if we go solely by GR. But what happens if QM is brought into the picture? Does that in anyway change anything about the way the Universe originated? That exactly is what Hawking asked himself. And it turns out that the man who said that the Universe *had* to originate in a singularity now said, the singularity is not really there because QM saves us! Thank God for QM! All those who had lost their sleep earlier, could now sleep in peace!! That story next.

Before I start describing what Hawking did, there are two things I must mention. First is that physicists abhor infinity because, roughly speaking, one does not know how to do mathematics when one has to start from infinity. Hawking demonstrated that if one brings QM into the picture, then one can, in principle, do all the algebra that one wants *without* encountering infinity. It is like the fireman wearing a damn good protection suit and going into a raging inferno to do his job. This is point number one. The second point is that it is not as if we know everything about how *exactly* the Universe got off to a start. Not at all. In fact, we know little about the precise state of the Universe at later times like 10^{-43} s or even as late as 10^{-30} s. Then why make a fuss about Hawking's work? Because he has given us the comforting news that *whatever* be the nature of the unified theory, QM would be there to protect and save the theorist from the ultimate disaster of having to deal with an infinity. So how does Hawking propose to set aside the infinity or get past it? That is what I shall discuss next.

We start with what our senses allow us to experience, the three dimensions of space and the flow of time. In relativity, space and time merge to become four-dimensional spacetime. Further, in QM many problems are routinely solved by resorting to the use of what is called *imaginary time.* Now how on earth can time be imaginary? It can't be but the mathematics becomes simpler with this transformation and that is why this trick is often resorted to. If you go back to ***Relativity*** (see, for example, Chapter 5) and look at some of the stuff there, you would find that imaginary time and space are on the same footing! So what these problem solvers would do is to carry out the algebra using imaginary time, which looks like 'space'. In effect, the use of imaginary time is like increasing the dimensionality of space by one. Once the job is done, the problem-solvers then wake us up from the anaesthesia by declaring that there are only three dimensions of real space and one of real time. In short, imaginary time is a tool, a mathematical tool to be more precise, as far as physics is concerned. There cannot be any logical or mathematical objection against using it.

What Hawking said was that time *truly* behaves like space in the ultimate quantum gravitational environment of the Big Bang. And when one takes account of this unusual behaviour of time, then the singularity problem is side-stepped. How? Let us see.

We start with Fig. 13.5, and imagine the Earth to be flat as in (a). In this case, if we go to the edge of the Earth, we are likely to 'fall off'! The flat Earth has two dimensions. Add one more dimension and make the Earth spherical. In this case, there is no question of falling off, as all of us know very well. OK, so what is the moral? Simply this. This disaster of falling off occurs because there is a singularity at the edge of the flat Earth—up to a point the Earth exists and beyond that point it ceases to. In the spherical Earth, however, no such disaster confronts us. In other words, by adding one more dimension of space, the singularity

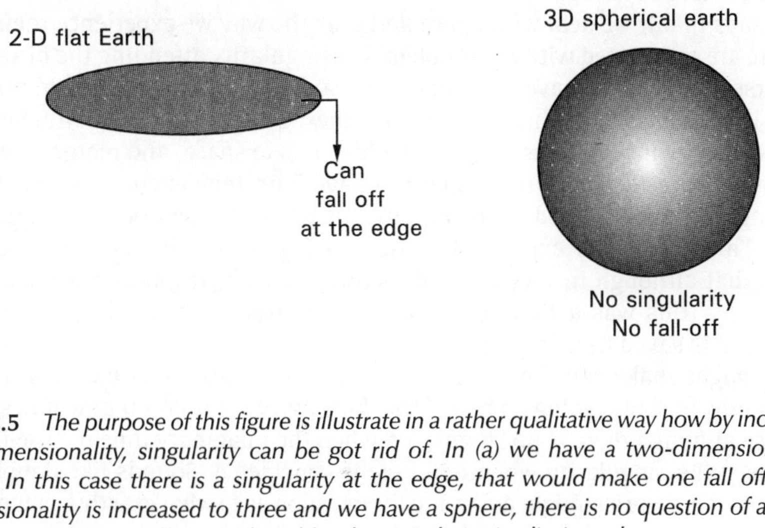

Fig. 13.5 *The purpose of this figure is illustrate in a rather qualitative way how by increasing the dimensionality, singularity can be got rid of. In (a) we have a two-dimensional flat world. In this case there is a singularity at the edge, that would make one fall off. If the dimensionality is increased to three and we have a sphere, there is no question of a fall off now. The disastrous effect produced by the singularity is eliminated.*

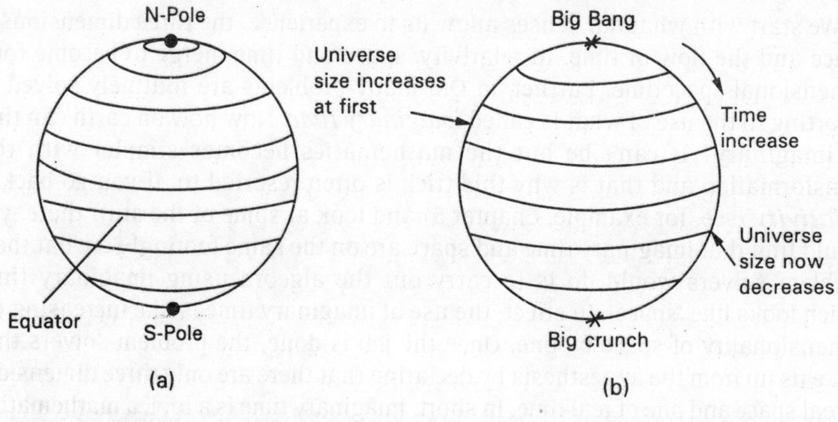

Fig. 13.6 *This is how Hawking explains the absence of singularity in the quantum picture and its presence in the classical picture. Let us start with the figure (a) on the left that shows schematically the Earth. Now a sphere (which is what we can approximate the Earth to) has no singularity. However, where the Earth is concerned, we have the poles. Consider the North Pole. At the exact N-Pole, there is no North, East or West; only South since any direction one moves is South. Now go just a little bit away from the N-Pole. This point has a North, East, West and South. There are thus clear distinctions, and this is what makes the N-Pole a singularity. It is in the same way that we try to understand the absence or presence of singularity with respect to the Universe. Quantum mechanically, with imaginary time and all that, there is no singularity. But once we go to the classical GR picture, we have a situation like in geography where poles are present. Here in (b), the Big Bang forms one 'pole', and is therefore a singularity. As one moves 'south' crossing various latitudes, time increases. Eventually, one reaches the S-Pole which represents the Big Crunch. What if the Universe never collapses? That scenario is discussed in the next figure.*

problem associated with the flat Earth has been taken care of. In essence, this is what Hawking does too.

He says that if we deal with space and time the way we experience them, then we sure are confronted with the problem of a singularity attending the birth of the Universe. Suppose we have no boundaries. We would then have no singularities —this is what happened when we went from the 2-D earth to the 3-D earth. Similarly, in this case we invoke imaginary time which is like space, and merge it with the traditional three dimensions of space to have four dimensions of space. Presto, the singularity associated with the birth of the Universe does not bother any more. This concept is explained further in Fig. 13.6. The long and short of this is that although in the classical picture the Universe *had* to originate in a singularity (this was a finding of Hawking himself), when the spice of QM is added one is saved that embarrassment.

You might shake your head vigorously and say, "I just don't like this. There is some funny business going on here!" Hawking agrees and says that would seem so because although there is no singularity when the imaginary time picture is used, there is still a singularity when real time is considered. Sounds like double talk, does it not? Is there a straight answer to the question as to what exactly is happening? Is there a singularity or is there not? Hawking gives this answer:

> This (Hawking's theory) might suggest that the so-called imaginary time is really the real time, and that what we call the real time is just a figment of our imagination. In real time (as we seem to understand it), the Universe has a beginning and an end at singularities that form a boundary to spacetime and at which the laws of science break down. But in (the so-called) imaginary time there are no boundaries or singularities. So maybe what we call the imaginary time is really more basic, and what we call real is just an idea that we invent to help us describe what we think the Universe is like. ....So it is meaningless to ask: Which is real, 'real' or 'imaginary' time? It is simply a matter of which is the more useful description.

What do you say to that?! I don't know whether you buy what the great man says, but in any case there are two things I must point out. At the time when Hawking developed this idea, people still believed that the Universe would one day start collapsing and finally end in a Big Crunch, the opposite of the Big Bang; in fact, I have myself discussed such scenarios in earlier volumes (see, for example, *A Hot Story*). But today, as we shall see in a later chapter, there are no takers for the story that one day our Universe would disappear in a Big Crunch. So, perhaps one should modify the illustration in Fig. 13.6 to what is shown in Fig. 13.7. This is my first point.

Next, we must not forget that we are nowhere near to describing properly the physics of the Universe at even Planck time. Therefore, there is no way we can

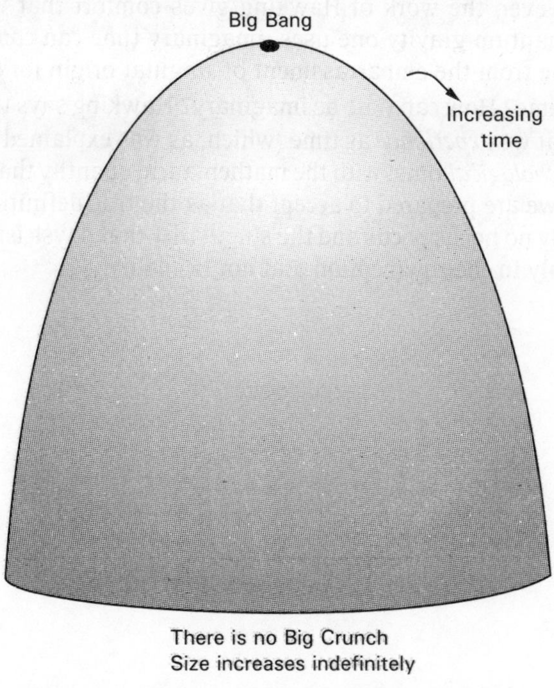

Fig. 13.7 *In this figure, what was represented as a sphere in Fig. 13.6 (b) is now shown as a semi-infinite paraboloid. There is thus only one "singularity", associated with Big Bang; there is none corresponding to the Big Crunch.*

discuss today precisely, even with imaginary time and all that, the exact situation at $t = 0$. But this we can possibly say, thanks to Hawking: If and when we can discuss the situation at $t = 0$, we need not be scared that we would have to confront a singularity. Imaginary time would be there to save us!

13.4 Summary

- Classical Cosmology of Friedmann, Le Maître etc., all suggest that the Universe started off in a singularity.
- Later work by Penrose and Hawking based strictly on general relativity provided a rigorous confirmation of this.
- But one cannot have infinity as the starting point! So what is the way out?
- Guided by his earlier work on the application of quantum mechanics to the physics of black holes, Hawking saw a way out.
- Essentially, Hawking showed that when one uses imaginary time, the number of space dimensions increases by one. With such an increase in spatial dimensionality, the singularity associated with the origin of the Universe disappears.
- True we do not have at present a proper theory for the Universe in its initial stages. However, the work of Hawking gives comfort that whatever be the theory of quantum gravity one uses, imaginary time can come to the rescue and save one from the embarrassment of singular origin for the Universe.
- Imaginary time? How can time be imaginary? Hawking says that we must not confuse what we *experience* as time (which, as was explained in ***A Hot Story***, is really *psychological* time) with the mathematical quantity that enters physical theories. If we are prepared to accept that as the true definition of time, then there is really no hocus pocus and the singularity that physicists are so terrified of, exists only in their perception and not in reality.

14

THE MYSTERY OF HIDDEN DIMENSIONS

We live in four-dimensional spacetime. However, theorists have been, for a number of reasons, formulating theories with more than four dimensions. Having done so, they then argued that these additional dimensions are 'hidden' from our view and experience. This chapter gives a brief overview of the question of why more dimensions, and what one does to 'hide' the extra dimensions.

14.1 Einstein's four-dimensional spacetime

We are all used to the three dimensions of space and are quite comfortable with the idea. Smaller dimensions such as two and one are of course no problem at all. But when about a century ago Einstein united space and time into four-dimensional spacetime, all but mathematicians had terrific problems, problems with visualisation that is. At the same time science fiction writers had a field day – they could concoct all kinds of scenarios.

During the past few years physicists have outdone science fiction writers by producing scenarios that these writers could never have dreamt of. And, unlike fiction, these scenarios are part of respectable scientific literature. Not only that, it seems that Nature is in fact structured that way! That is what I shall briefly deal with in this chapter.

It all started soon after Einstein published his historic work on general relativity. In this, Einstein said (effectively): "Space and time are not distinctive and separate though our sense perception makes us feel that way. They are united as spacetime, which is four-dimensional. This four-dimensional spacetime can be flat, spherical or hyperbolic. If matter is present, it distorts this spacetime, producing bumps and dimples. Particles, and for that matter light, travel in this spacetime along paths that have the least length. Thanks to the bumps and dimples, it appears that matter is attracted by matter and light too appears to feel a similar attraction to matter. It is this bending that we interpret as attraction due to gravity." By the way, I do hope you recall the famous experiment about bending of starlight that propelled Einstein into celebrity status (see ***Relativity***).

14.2 Kaluza makes it five-dimensional

Around 1919–1920, a young scientist named Kaluza in USSR read Einstein's classic work on general relativity. He then said to himself: "Einstein says that gravity is just a manifestation of curvature of 4-D spacetime. Is it possible to understand the electromagnetic force also in the same manner?" After pouring over the matter, Kaluza came to the conclusion that if Einstein's 4-D space was made 5-D, then both gravity and electromagnetic forces could perhaps be regarded as consequences of curvature of this higher dimensional space. Kaluza wrote a paper based on his findings, and, as Bose was to do some years later, sent the manuscript to the great master. The story of what happened thereafter is told in BOX 14.1.

BOX 14.1

In the beginning of 1919, Kaluza wrote to Einstein (as Bose was to do a few years later) submitting a copy of his manuscript. In reply, Einstein wrote on 21 April, 1919:

> The idea that this (representation of the electric field) can be achieved through a five-dimensional cylinder-world never occurred to me and would seem to be altogether new. I like your idea at first sight very much ....If no apparently fatal and irreparable objection occurs to me when reading your detailed exposition, I will be pleased to present your paper on the subject to the academy here.

Just one week later, that is on 28th April 1919, Einstein wrote:

> I have read through your paper and find it really interesting. Nowhere, so far, can I see an impossibility. On the other hand, I have to admit that the arguments brought forward so far do not appear convincing enough.

Cold water for Kaluza! Two years later, however, Einstein changed his mind and wrote:

> I am having second thoughts about having restrained you from publishing your idea on a unification of gravitation and electricity two years ago. Your approach in any case seems to have more to it than the one by Weyl. If you wish, I shall present your paper to the academy after all, provided you send it to me.

And that is how Kaluza's paper finally got to be published, after an unnecessary delay of two years. The flip side is that unknown Kaluza's paper was rated higher by Einstein than the paper by Hermann Weyl, already a big name!

In essence, what Kaluza said was this: "Einstein says spacetime is four-dimensional. If matter is present, then this spacetime becomes warped. In fact, we simply interpret this warping as the presence of matter. Gravitational waves emitted by moving matter are ripples propagating in warped 4-D. Now, light follows the shortest path or the geodesic in curved space; imagining 3-D space to be flat, we say light travels in straight lines. Thinking in terms of 4-D, when there is no matter, light would follow the shortest path in spacetime. If this 4-D is spherical say, then light would follow the great circle in this spherical 4-D spacetime. If matter is present, then the trajectory would have to go along a bump near the place where matter is located. This, we say, is due to the attraction of light by matter. What happens to light also happens to gravitational waves. All this is what we know from Einstein's theory involving 4-D. Suppose now we have five dimensions instead of just four. Then, according to me, just as matter is a bump and gravitational waves are ripples in 4-D, charges are bumps and electromagnetic waves are ripples in 5-D. In short, electromagnetic waves would appear to get deflected when there are dimples and bumps due to both masses and charges."

I admit I have grossly oversimplified Kaluza's theory, but for our purposes this suffices. Many questions now arise such as: "Is this just a geometric idea or can this theory explain the presence of different kinds of particles, their masses, charges and so forth?" Obviously, Kaluza's theory could not do any of these things, and his is thus essentially a *geometric* way of looking at Nature – no doubt an interesting way one must admit. In that case, why talk about it at all, in this day and age? That I shall explain presently.

I must also stress that Kaluza did *not* unify gravity and the electromagnetic force. In fact Einstein tried to do precisely that all his life but never succeeded. That was because Nature had paired the electromagnetic force with the weak force and *not* the gravitational force, as Einstein had imagined it was. Nature had bowled a very clever and highly concealed googly, and Einstein did not read it! I mean, it is natural to suppose that gravity is the brother of the electromagnetic force; they are absolutely similar in terms of the inverse square law; how could they not be natural partners? So Einstein thought. But Nature does not work that way. It took Fermi, and then the Salam–Weinberg–Glashow trio to find the correct partner of the electromagnetic force. That story has been narrated in Volume 1. All I am trying to tell you presently is that Kaluza did not unify gravity and the electromagnetic force but gave instead a convenient way of visualising electromagnetic and gravitational forces in *geometric* terms, via a 5-D space.

14.3 Klein adds quantum mechanics to Kaluza's picture

Both Einstein's theory and Kaluza's extension are classical. When quantum mechanics appeared on the scene a few years later, Oscar Klein in Sweden asked himself how these conclusions of Kaluza would get modified by the application of the ideas of quantum mechanics. Klein wrote a 5-D version of the Schrodinger equation and showed that the solutions of this equation can be interpreted as waves moving in 4-D under the combined action of gravity and electromagnetic forces. This extended type of theory is referred to as *Kaluza–Klein theory.* Thanks to Klein, people could now play a new game. Just as people tried to understand the quantum aspects of the atom by imagining different types of de Broglie waves wrapped around the electron orbit (see ***WATT***), people now started asking somewhat similar questions.

As a prelude to an explanation of this type of analysis, I must mention that many wondered how it was that we do not experience this fifth dimension. An answer was evolved based on the following argument: 'Suppose at every point in the 4-D spacetime we have a circle. This circle would represent the

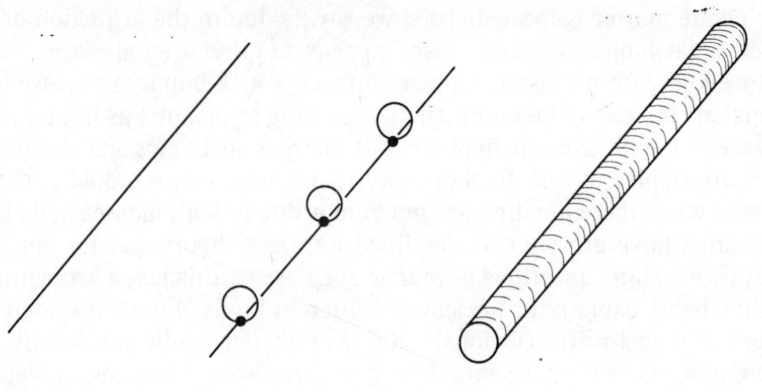

Fig. 14.1 *This figure illustrates how one is to understand compactification. We start with (a), which shows a line. Suppose we place a circle at every point on the line. In (b) are shown some of these circles. If circles are placed everywhere, we get a tube as in (c). As is evident, the surface of the resulting cylinder is two-dimensional. If now we compactify the circle so that its diameter becomes very small, then the tube/cylinder begins to look effectively like a line. That is the general spirit of compactification. Space does have more dimensions than four; but thanks to heavy compactification it appears to have just 3 + 1 dimensions, that is all.*

extension of 4-D into the fifth dimension at that point. Let R be the radius of the circle so erected. If R is very small, then we may not experience it at all."

The following analogy may be useful in this context. Imagine a long tube. The tube may be obtained by starting with a long line and adding a circle at every point. Whereas the line has only one dimension, the tube has two; this you will agree to. Now suppose the diameter of the circle becomes very small, say about an angstrom. Technically, we still have a tube but an ant crawling on the tube would think it is just a geometric line of one dimension (see Fig. 14.1).

People interested in quantum mechanics now started fitting waves of different wavelengths around the circle of radius R in the fifth dimension (similar to what people did with de Broglie waves and electron orbits). This led to some idea of the particles and their masses; but the masses came out to be too large and did not have much to do with reality, whereupon people naturally lost interest in the Kaluza–Klein theory.

14.4 Compactification and modern versions of theories with hidden dimensions

If people had lost interest in Kaluza–Klein kind of theories long ago, then why on earth am I talking about an idea that went up like a lead balloon as they say? Because I want to introduce you to a new word, *compactification.* This is a word you will hear more often later. What this means is the shrinking of a dimension in a theory that is over and above the four dimensions that we are used to, and the acceptance of the latter as a reality of experience. In Kaluza's theory, the fifth dimension was sought to be compactified by making it a circle of very small radius. We will have more of this shortly and in later chapters also.

Having introduced the word compactification, let me get back to physics, current physics in fact. When we talk of Kaluza–Klein, we are talking of the twenties of the twentieth century. Years passed and the Kaluza–Klein theory was generally forgotten. Then one fine day, someone suddenly woke up and said, "Hey, when Kaluza wrote his paper no one knew about the weak force and the strong force. Today, we have four basic forces instead of two. How many dimensions are required to describe these?"

Here the game becomes complex, thanks to all kinds of internal symmetries that have already been mentioned in Volume 1. As a result, one cannot just say, "Kaluza had two forces, therefore he needed five dimensions. I have two more, so I need seven dimensions," or something like that.

Agreed that one cannot make a naïve extension like that. So how many dimensions are *actually* needed if we are to have a Kaluza–Klein (KK) type of theory for all the four forces? This question has received a lot of attention. The first thing that people noted was the KK type of theory allowed room for only bosons. Remember that graviton (the quantum of the gravitational field) is a boson and so also is the photon, the quantum of the electromagnetic field. Moreover, in the Kaluza–Klein analysis there is no explicit reference to the fact that charged particles like electrons are fermions. When we include the weak and the strong forces, we can no longer avoid such explicit reference to fermions, which means that the KK theory cannot be blindly extended. This is the first point. Next, additional dimensions should not be postulated or invoked but *must emerge naturally.* In a sense, this is what happened when people began to investigate supergravity and String Theory.

These people who were working with supergravity and super string theories involving several extra dimensions soon began to ask, "Whatever happens to these extra dimensions? We obviously cannot have ten or eleven dimensions since we live in a four-dimensional world. There must be some way of packing off these extra dimensions once they have done their job and played their role." Thanks to earlier history associated with Kaluza–Klein theories, physicists naturally began to think in terms of compactifying the extra dimensions so that one finally ends up with three dimensions of space and one of time. In the case of the most popular version of supergravity theory, for example, there were 11-dimensions to start with, which automatically means that seven extra dimensions have got to be compactified.

Several questions arise immediately. How exactly should the compactification be done, and what should be extent of compactification (I mean what should be the value of R in the example discussed earlier?). Further, at what point during the evolution of the Universe does compactification occur? Also, does the size of the compactified dimension remain constant or does it change with time? This is where some of the work done earlier by mathematicians helped greatly. Mathematicians have no difficulty

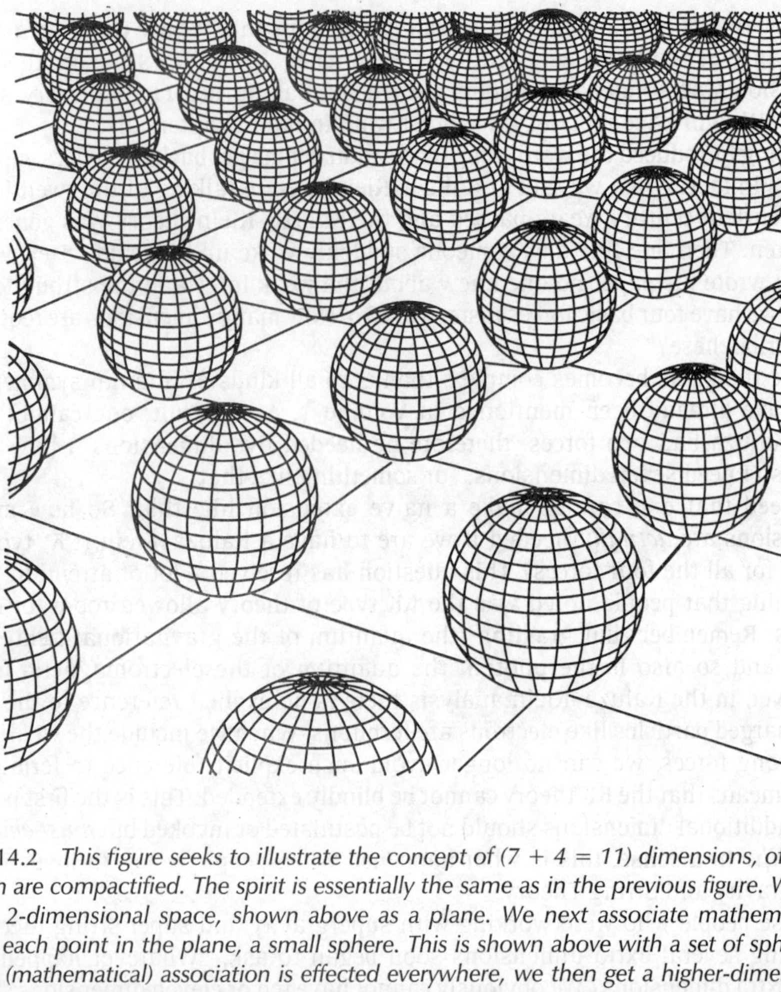

Fig. 14.2 *This figure seeks to illustrate the concept of (7 + 4 = 11) dimensions, of which seven are compactified. The spirit is essentially the same as in the previous figure. We start with 2-dimensional space, shown above as a plane. We next associate mathematically with each point in the plane, a small sphere. This is shown above with a set of spheres. If such (mathematical) association is effected everywhere, we then get a higher-dimensional space. In supergravity, one has to start with 11 dimensions. This space is visualised as (3+1) dimensions extended by seven extra dimensions. This extension can be done in various ways, by hyperbolic spaces, by spherical spaces and various combinations thereof. It turns out that extension by seven-sphere is most elegant and physically appealing. The seven-sphere is like the sphere shown above. The usual 4-D spacetime results when the seven spheres are all compacted.*

in working with spaces of any arbitrary dimension because they don't have hang-ups about visualisation that bother us physicists so much. Thus, a long time ago they have studied all sorts of things about all sorts of spaces of all sorts of dimensions! Drawing upon this fund of knowledge, supergravity physicists said, "The extra seven dimensions must be shrunk into a seven-sphere, that is, at every point in 4-D spacetime we must have a seven-dimensional sphere." This idea is schematically shown in Fig. 14.2. Note that there are many different ways of compactifying seven dimensions but the seven-sphere idea seemed most elegant and closest to physical reality.

When it came to String Theory, it was a 10-dimensional theory with compactification of six extra dimensions that seemed worth examining further. By the way, all this was during the period of what is sometimes called the First String Revolution. Subsequently, we have had the Second String Revolution, and things have evolved a lot since the early days. Higher dimensions are still around but the story is more complex and we shall take a glimpse of all that in a later chapter. Meanwhile, it is generally believed that some kind of a minimisation principle possibly drives compactification. It is also taken for granted that once a dimension is compactified, its size remains constant even though spacetime itself may expand.

14. 5 Summary

- It now seems inevitable that for getting a complete theory in 4-D, we would have to start with a 'parent' theory in higher dimensions. This higher dimension could be 10, 11, or even higher. The matter is not settled yet.
- After the higher-dimensional theory is fully worked out, we must descend to 4-D to make contact with the *real* world. This calls for compactification of the extra dimensions in a suitable manner and to a suitable size.
- Size of the compactified dimension? No one knows for sure but most guesses are for very small distances, of the order of Planck length.
- Will all compactified dimensions have the same size or would the sizes be different? No one knows.
- When exactly does compactification occur during the evolution of the Universe? Possibly during the Planck era.
- Does the size of the compactified dimension stay constant or vary with time? Almost everyone is in agreement that these compactified sizes remain constant from the moment compactification occurs.

Although the basic concepts appear simple, the technical details are definitely not. As we shall see in a later chapter, when compactification was attempted in super string theories, millions of 4-D candidates emerged! Which of these actually described our real world? Wait some more, and after we have acquired the necessary background I shall tell you how physicists have been dealing with that particular box of Pandora!

15

IS SOMETHING MISSING?

There is now abundant evidence that luminous matter alone does not make up all the matter in the Universe. It would seem that there is something else, called dark matter. This chapter discusses briefly the evidence for such non-luminous matter, considering both the macroscopic as well microscopic contributors.

15.1 Introduction

For thousands of years astronomy has relied totally on the visibility of cosmic objects. To study the object one had to physically see it, at least with the eye, if not with an optical telescope. Later, other types of telescopes like the radio telescope were added, all the way to gamma-ray telescope. Does this mean that objects cannot or do not exist that cannot be seen at all by any type of telescope? Of course they can and the black hole is a classic example. However, serious attention was given to the existence of 'invisible' or non-luminous matter in the context of Cosmology only when people began to earnestly study the future of the Universe.

As already mentioned, the future of the Universe is determined by the value of Ω (see Chapter 6). Many years ago it was believed that matter associated with stars and galaxies alone contributed to Ω . (Not all stars have planets around them, and if our solar system is an example, then 99 % of the matter in a planetary system is resident in the central star. In other words, for this exercise we can safely ignore the planets and just count the stars in the galaxies. And that has been done, roughly of course.) When Ω was estimated in this manner, it turned out to have a rather low value, approximately around 0.2. There was no way in which this low value could be enhanced significantly by saying, "Oh no, our count is very rough!" True the galaxy and star count is rough, but the accuracy is good enough for the present purposes. Note that even if a 100 % error is made, we come up with only a value of 0.4 for Ω . On the other hand, other recent studies (not involving star and galaxy counts) strongly suggest that Ω must be close to 1 (more about this in the next chapter). Now there is good reason to believe that the galaxy count is correct. If that is so, then how come Ω is close to 1? I mean, there does not appear to be enough matter in the Universe, and yet Ω comes out to be close to 1. How to reconcile this difference? It is in this context that the entire issue of *dark matter* became very important.

15.2 Evidence for dark matter in galaxies

Galaxies are luminous. The name dark matter is given to matter that is *not* luminous. When it was discovered that $\Omega \sim 1$, it became clear that there must be a lot of non-luminous matter out there. If so, how much and of what type? Those are the questions we shall briefly deal with presently. You might ask: "If matter is not luminous, how can we know it is present?" Well, by its gravitational effect, among other things.

Let me now describe one astrophysical study that provided the first clear evidence of the existence of non-luminous matter. This study was carried out by astronomer Vera Rubin on a large number of spiral galaxies. Now it is generally observed that the luminosity of a spiral galaxy drops off from the centre to the periphery. This seems to suggest that there is more mass at the centre than the periphery. But what if there is also matter that is not luminous, or dark matter if you will? What is one to expect in such a case? This is what Vera Rubin investigated. Her focus was on the orbital velocity within the galaxy at various distances from the centre; so let us first try to understand how we expect this orbital velocity to vary in a normal galaxy.

Let us start with our solar system where we have the Sun at the centre and the planets in various orbits around it. The striking thing about the solar system is that more than 99% of the mass of the solar system is concentrated in the Sun! In such a case, the planets show a $(1/r^2)$ variation of the orbital velocity, a Keplarian pattern. Is this what we should expect for a galaxy? Let us examine. A spiral galaxy has a core or a nucleus with arms emerging out of it. Suppose for a moment we regard the core as a solid sphere. As you know from first year physics, inside the sphere the orbital velocity will increase with r. If now most of the mass of the galaxy is concentrated in the core, then we expect the overall curve to be a mixture of the two types mentioned above, something like a linear increase at small distances and a Keplarian sort of decrease beyond the core. The above ideas are pictorially summarised in Fig. 15.1

Over now to the careful study made by Vera Rubin. She surveyed over sixty galaxies, and the principle behind those investigations is illustrated in Fig. 15.2. She found that in many of the galaxies she had investigated, the orbital velocity as function of distance departed significantly from the expected behaviour as shown in Fig. 15.3(a). This clearly meant that mass did not decrease with distance in the same manner as the luminosity of a normal galaxy. There was a lot of non-luminous mass hiding, and the question was, where?

Many scenarios for the hidden dark mass were envisaged and one of these is the halo model. In this it is assumed that the galaxy is embedded in a 'halo' of dark matter that extends well beyond the visible limits of the galactic disc. If such a halo did exist, then the mass associated with this halo could alter the orbital velocity curve. This is how the dark matter is expected to show itself up, though non-luminous.

Figure 15.4 compares what may be expected in the standard type of spiral galaxy and one embedded in a dark halo. The orbital velocity behaviour observed by Vera Rubin was, in many cases, in accord with the orbital velocity

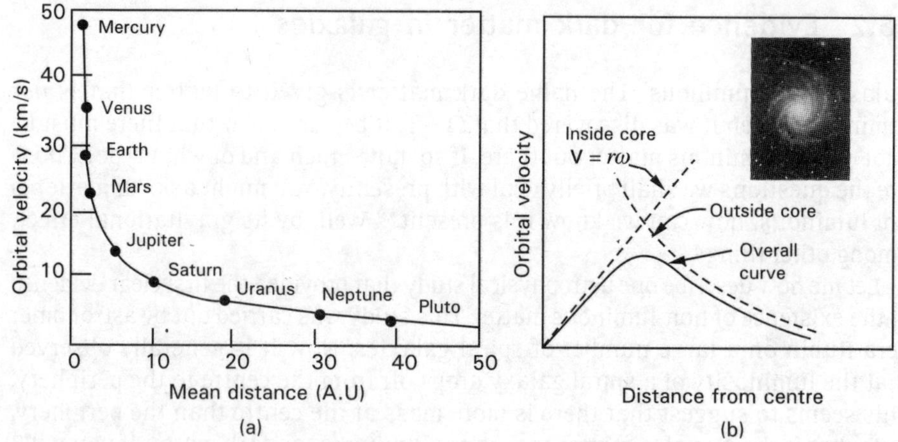

Fig. 15.1 *This figure compares the variation of orbital velocity with distance in the solar system (a) and the variation of the same roughly expected in a spiral galaxy (without any dark matter). (b) In the solar system, where over 99 % of the mass is concentrated entirely in the Sun, the orbital velocity varies in a strictly Keplarian fashion, i.e., as $(1/r^2)$ as shown in (a). To understand the velocity variation in a spiral galaxy, we first note that in the core it would increase with r while outside the core it would decrease with r, assuming that mass concentration is in the core. The overall curve would thus be a composite of the linear increase with r within the core and a $(1/r^2)$ type of behaviour outside the core as shown in (b).*

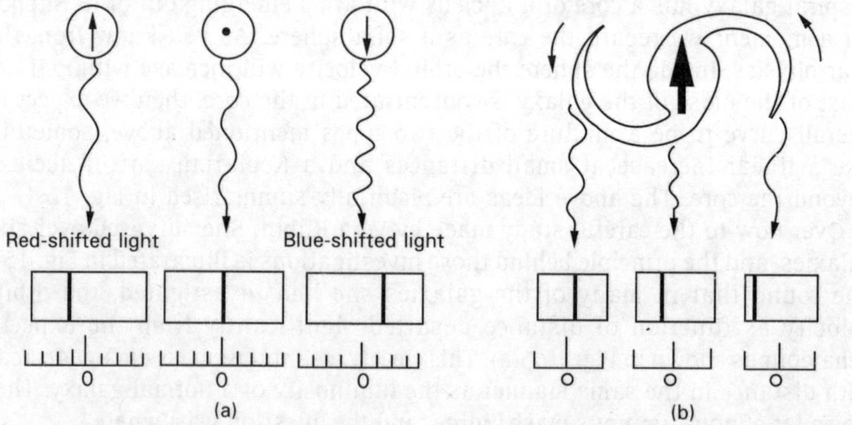

Fig. 15.2 *This figure illustrates the principle behind the Vera Rubin studies. In (a) is illustrated the basics of the familiar Doppler principle. The frequency of light emitted by a moving source is blue- or red-shifted with respect to the frequency of the stationary source, depending on whether the source is moving towards or away from the observer. Now look at (b). Here the galaxy itself is moving away from the observer; and so there is a basic red shift. However, with respect to this basic motion of the galaxy as a whole, one arm is moving towards and the other is moving away from the observer. This would produce corresponding redshifts as shown.*

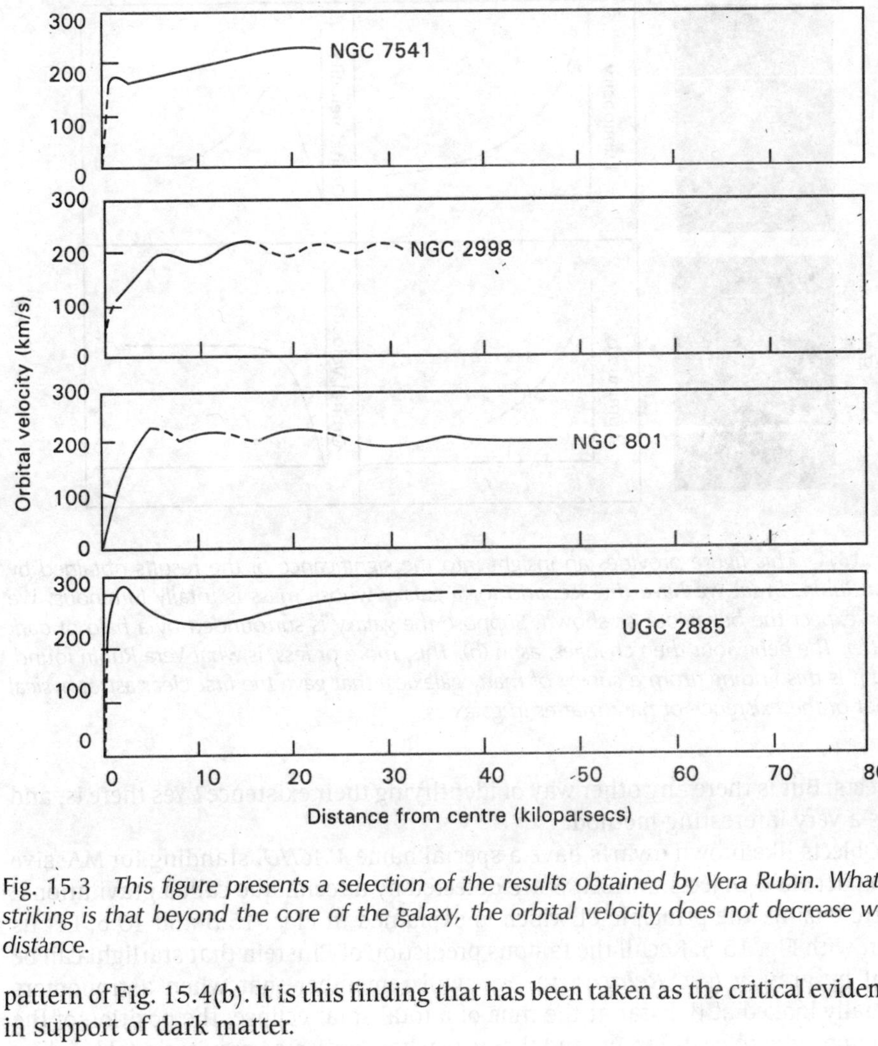

Fig. 15.3 *This figure presents a selection of the results obtained by Vera Rubin. What is striking is that beyond the core of the galaxy, the orbital velocity does not decrease with distance.*

pattern of Fig. 15.4(b). It is this finding that has been taken as the critical evidence in support of dark matter.

15.3 MACHOs as dark matter candidate

OK, so we possibly have dark matter hanging around as a halo close to galaxies. Can we say more? There are many suggestions, one of which is that dark matter is made up of may be either black holes or less exotic objects known as *brown dwarfs*. A brown dwarf is a celestial object that is anywhere between 10 to 80 times heavier than Jupiter. Brown dwarfs form the same way as stars do, condensing from a cloud of hydrogen. However, unlike stars, they do not accumulate enough mass to generate by gravitational contraction the very high temperatures needed to kick off thermonuclear reactions at the core. Instead, brown dwarfs glow the same way as do giant gassy planets like Jupiter, heating up to a moderate extent. Since these brown dwarfs are very far away, we cannot see them as luminous

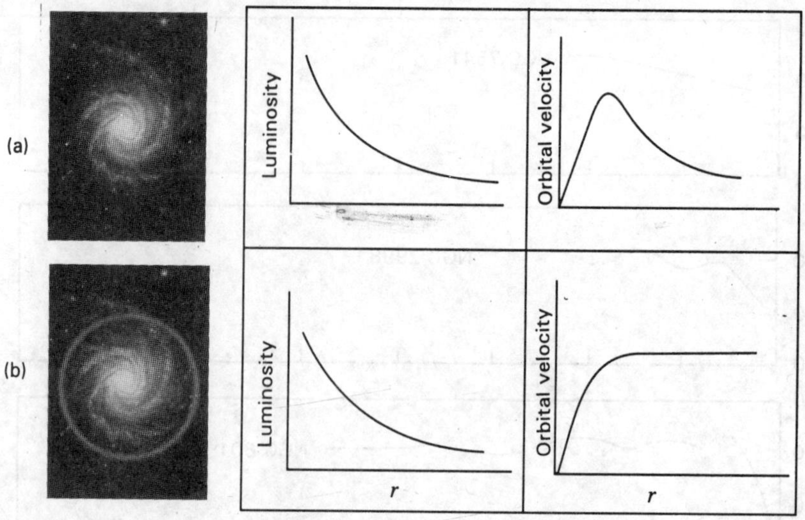

Fig. 15.4 *This figure provides an insight into the significance of the results obtained by Vera Rubin. In (a) we have the scenario for a galaxy whose mass is totally luminous. We then expect the behaviour as shown. Suppose the galaxy is surrounded by a halo of dark matter. The behaviour then changes, as in (b). This, more or less, is what Vera Rubin found, and it is this finding (from a survey of many galaxies) that gave the first clear astrophysical proof of the existence of dark matter in galaxies.*

objects. But is there any other way of identifying their existence? Yes there is, and it is a very interesting method.

Objects like brown dwarfs have a special name *MACHO*, standing for MAssive Compact Halo Objects. MACHOS are detected by a technique called gravitational microlensing, the principle of which is explained in Figs. 15.5 and 15.6. Let us start with Fig. 15.5. Recall the famous prediction of Einstein that starlight can be bent by gravity (see ***Relativity***). You will remember that when astronomers actually looked at the star at the time of a total solar eclipse, the position of the star appeared to be different; and that was what confirmed gravitational bending of light. In the seventies of the twentieth century, people began to develop the idea that massive objects in the cosmos could in some way act like a lens, a gravitational lens really. As someone put it, gravitational lensing is geometric optics done the general relativity way! Soon after, theoretical speculations about gravitational lensing were first aired, an actual lensing event was detected in 1979; thereafter, gravitational lensing has been taken quite seriously. The lensing action of dense matter is actually quite complicated, depending as it does on the distribution of matter, the motion of the object, the lens and the observer, and so on. Obviously, I cannot go into all that here. So let me instead offer a hand-waving version of this lensing business, with specific reference to MACHOs.

As you know, in optics we have thin as well as thick lens. A MACHO is supposed to act like a thin gravitational lens, and its action is referred to as microlensing. Let us suppose there is a star far away and that we look at it with a telescope. Say the

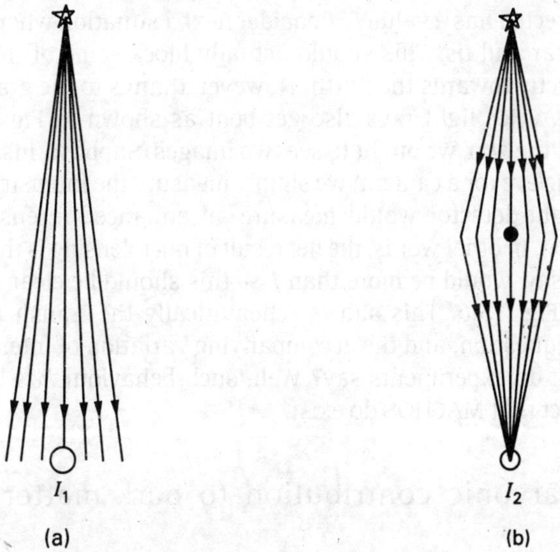

Fig. 15.5 *This figure illustrates the meaning of microlensing. In (a) we have a star whose light reaches the Earth without any obstruction. In (b), a MACHO intervenes. Thanks to the gravitational pull of the MACHO, light beams that would otherwise have gone away from the Earth are now pulled in. This causes a net enhancement of the intensity of starlight, as explained in the text. That is what microlensing is all about, in essence.*

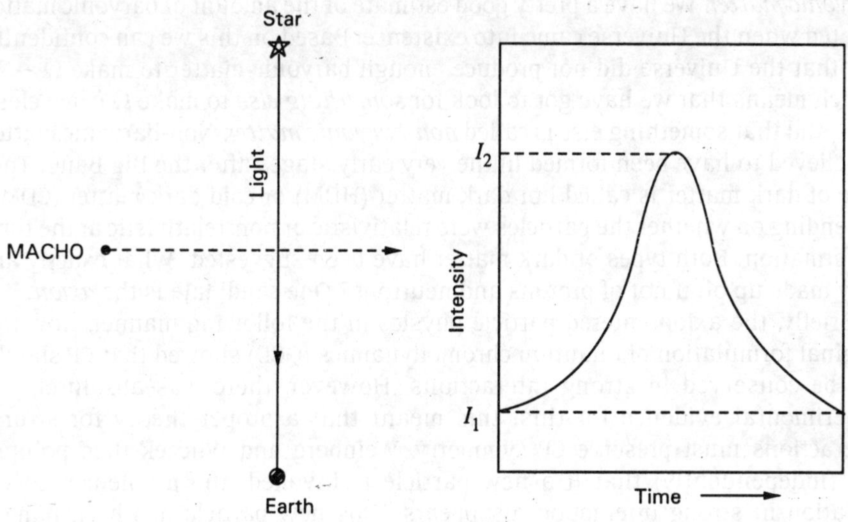

Fig. 15.6 *This figure sketches the expected variation of intensity of starlight as a MACHO transits across the star in question. Several cases of such intensity enhancement have been detected, confirming the existence of MACHOs.*

light intensity detected has a value I_1. Consider next a situation when a MACHO sits in between the star and us. This should actually block some of the light coming from the star directly towards the Earth. However, thanks to the gravitational pull exerted by the MACHO, light rays also get bent as shown in Fig. 15.5(b). If we actually looked at the star, we ought to see two images. Suppose, instead of looking at the star with the eye (or a camera) we simply measure the intensity of light at the eyepiece. This light-detector would measure an enhanced intensity due to the microlensing action. In other words, the net result of microlensing is that the intensity now observed, I_2 say, would be more than I_1 —this should be clear from Fig. 15.5. We now turn to Fig. 15.6. This shows schematically the transit of the MACHO across the star in question, and the accompanying variation of intensity. All this is speculation; what do experiments say? Well, such behaviour *has* been observed, confirming the fact that MACHOS do exist.

15.4 Non-baryonic contribution to dark matter

OK, so MACHOS exist; do they make up all the dark matter possible in the Universe or are there other candidates? It looks like there are because MACHOS cannot by themselves contribute enough matter to make $\Omega \sim 1$. How do we know? Well, there are pretty good estimates of how much primordial hydrogen was formed, say about few minutes after the Big Bang (recall the discussion in Chapter 9). Now MACHOS are made up of ordinary atoms and molecules as is Jupiter, that is to say they are essentially made up protons and neutrons or *baryonic matter*. We have a pretty good estimate of the amount of baryonic matter created when the Universe came into existence. Based on this we can confidently say that the Universe did not produce enough baryonic matter to make $\Omega \sim 1$. Which means that we have got to look for *something else* to make Ω come close to 1. And that something else is called *non-baryonic matter*. Non-baryonic matter is believed to have been formed in the very early stages after the Big Bang. This type of dark matter is called hot dark matter (HDM) or cold dark matter (CDM), depending on whether the particles were relativistic or non-relativistic at the time of formation. Both types of dark matter have been suggested. What exactly are they made up of, if not of protons and neutrons? One candidate is the *axion*.

Briefly, the axion entered particle physics in the following manner. Now the original formulation of quantum chromodynamics (QCD) showed that CP should not be conserved in strong interactions. However, there was absolutely no experimental evidence for this; this meant that a proper theory for strong interactions must preserve CP symmetry. Weinberg and Wilczek then pointed out (independently) that if a new particle is invoked, the problem with CP violation in strong interaction disappears. This new particle has been named axion. It is believed that most of the axions in the Universe (if they exist!) would have been produced when the Universe was very hot ($T = 1$ GeV). They are produced in interactions like:

nucleon + pion → nucleon + axion,

gamma + quark → quark + axion.

Many ideas have been proposed to suggest how axions could pervade the Universe and contribute to dark matter, especially by association with galactic halos. Experiments to detect the axion have been suggested but as of now there is no clear proof of the existence of the axion.

Another non-baryonic candidate is the *WIMP*, the acronym for Weakly Interacting Massive Particles. Go back to SUSY and recall the supersymmetric particles the bino, the wino and the higgsino (partners of the B meson, the W particle and the Higgs particle respectively). In one version of a supersymmetric theory there is a particle called the *neutralino* that is a liner combination of the bino, the wino and the higgsino. The WIMP is nothing but the neutralino. Its mass is expected to be in the range of tens to hundreds of GeV, and it is expected to interact weakly with matter. It too is a candidate for dark matter. At present the WIMP is just an idea; it is yet to be detected.

OK, there you have it, a quick survey of dark matter. People became excited about dark matter because they thought that dark matter would make up the shortfall in Ω, bringing its value close to 1. It did not. People are now agreed there is definitely dark matter in the Universe but not enough of it to make $\Omega \sim 1$. So how does one jack up the value of Ω? You know how that has been done? By reviving the good old cosmological constant Λ that Einstein sought to bury! Thus, it is now quite fashionable to talk of Λ! The wheel has come full circle; that story is picked up in the next chapter.

16

WHERE WILL IT ALL END?

This chapter deals with the ultimate fate of the Universe. It turns out that this ultimate fate depends largely on the value of Ω. Based on the value of Ω then available, it was speculated a few decades ago that the Universe might even end in a Big Crunch. Subsequent developments have ruled out that possibility. Currently, most people feel that the Universe will asymptotically reach a constant size. There are, however, some suggestions based on data from supernova that the Universe might rather expand indefinitely into the future. These questions are discussed in this chapter.

16.1 A refresher on Ω

Let us suppose that the Universe was born in a Big Bang, with inflation and all that; for a moment we shall forget the complications that flow from Linde's chaotic inflation and self-generating cosmos. We also know that since birth, the Universe has been expanding; in fact, it is expanding right now. But what about the future? What would be the fate of the Universe? Can physics say anything about that? That story revolves largely around a famous parameter of Cosmology, namely Ω. (If Linde's model applies, then Ω would presumably be that relevant to our Universe.)

To understand what Ω really implies, we recall from Fig. 6.4 that there are three distinct scenarios possible for the future of the Universe. This is what may be called a 'first look', obtained by considering just the *geometry* of the Universe, or spacetime if you prefer it that way. We shall now try to understand better the meaning and true nature of Ω. For this purpose, we go back first to Hubble's law and the Hubble constant.

In Chapter 6, I merely related this constant to the experimental finding of Hubble, besides quoting the current value of the constant denoted by H_0. What does Cosmology have to say about all this? Well, if one starts with the Robertson metric (that I mentioned in Chapter 6) for the Universe and works one's way through, one obtains a simple relation of the form

$$\left\{\frac{R(t)}{\dot{R}(t)}\right\} = H(t).$$

This relation essentially defines the Hubble constant at time t (often referred to as an epoch). It is customary to denote the current epoch by t_0, in which case we have

$$H(t_0) \equiv H_0$$

which was the constant I introduced you to in Chapter 6. We next define for every epoch t, a critical density $\rho_c(t)$ by

$$\rho_c(t) = 3\left[\frac{H^2(t)}{8\pi G}\right].$$

In particular, for the current epoch t_0, this gives,

$$\rho_c(t_0) = 3\left[\frac{H_0^2}{8\pi G}\right].$$

In Cosmology, the dimensionless parameter Ω is defined by

$$\Omega = \frac{\rho(t_0)}{\rho_c(t_0)},$$

that is, the ratio of the current actual density $\rho(t_0)$ to the current critical density introduced above. It is this quantity that is usually denoted as Ω and defined simply as (ρ/ρ_c). The time t_0 is not always made explicit. Experts understand all this of course, leaving us lesser mortals to figure out the fine print!

At this stage you might begin wondering about all this (as I did too) since Ω as defined above is for the present epoch. You might say: "Listen, there appears to be some funny business going on here. When you introduced Friedmann's theories in Chapter 6, you drew many graphs, and said these applied to different values for Ω. That would seem to imply that Ω does not vary with time. But in the above, you have defined Ω with respect to quantities pertaining to the present epoch. How then can I use this definition of Ω in understanding Friedmann's curves? After all, the density would change with time, and so would the Hubble constant, as per your definition."

Yes, all these points are well taken. But you see, the beauty is that for every instant of time, there are: (i) a particular value for the Hubble constant, (ii) a particular value for the density of the Universe, and (iii) a particular critical density. If we take these factors into proper consideration (and this comes out from the details that we skip), then it turns out that where Ω is concerned, it really does not matter what the value of t is! For numerical work and model analysis, however, we use the 'current' value of Ω, because from experiment we have both a number for H_0 and an estimate for $\rho(t_0)$.

With all this background we can safely treat Ω as a constant (although for practical purposes we always mean $\Omega(t_0)$, since the numbers used are from today's experiments). Let us now go back to the implications of the various values possible for Ω. As already noted earlier, the Universe expands without limit if $\Omega < 1$, and becomes closed when $\Omega > 1$. For the critical value $\Omega = 1$, the Universe is flat; that happens when $\rho = \rho_c$.

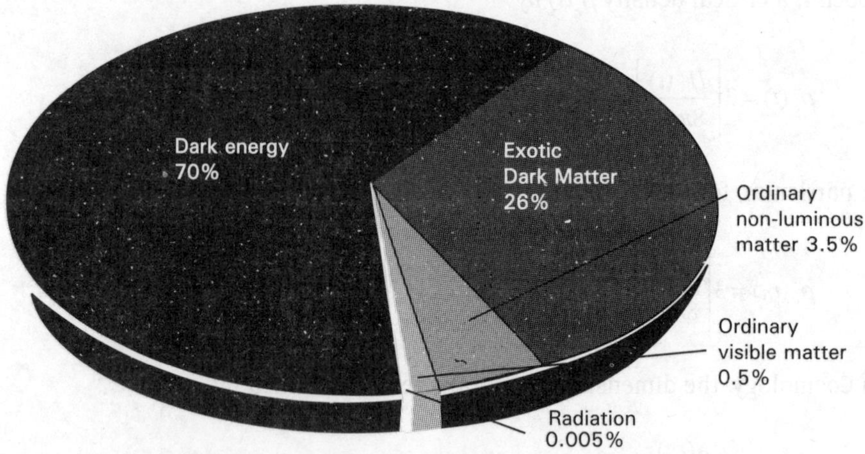

Fig. 16.1 *This pie chart is essentially an approximate 'census' report of the cosmos (the percentages do not exactly add up to 100 because of rounding off). Many years ago, people believed that the Universe was filled only with radiation and matter, with matter being made up of galaxies, which in turn were made up of stars and perhaps their planets. The view now is very different, as the above figure shows. Ordinary visible matter forms a very small part, and there is much more non-visible matter. In addition, there is the so-called non-baryonic or exotic dark matter explicitly invoked by particle physicists investigating Cosmology. But even all these added together do not bring the value of Ω close to 1. Current evidence is that Ω is close to 1. So how to close the gap? Bring in dark energy, which, it turns out, has the lion's share of Ω! Later we shall consider attempts to link this dark energy to Einstein's Λ.*

16.2 Cosmic census

One thing becomes immediately clear from all that has been said above. If we want to know about the future of the Universe, we have to first of all determine ρ_m, the density of matter, and the contribution it makes to the value of Ω. I have touched upon all this in the previous chapter on dark matter. We noted there that after much work, people found that even with dark matter the value of Ω fell far short of 1; clearly, something *more* was needed to make Ω approach the value 1. This is where the story becomes very interesting; let me beginit by first drawing your attention to Fig. 16.1 which presents what might be called a cosmic census report.

16.3 Ripples in CMB

To take the story further, I must now tell you why people believe that Ω must be close to 1. Cut back to the cosmic microwave background (CMB). Earlier, I told you in Chapter 6 that that following the epoch-making discovery of Penzias and Wilson, the CMB was studied fairly extensively and two facts were established. First, the temperature of the radiation was found to be 2.728 ± 0.002 K. Second, the radiation was found to be isotropic, that is the radiation coming from different directions in the sky was found to be the same to within 1 part in 10^4. After a while astronomers said, "Let us look more closely at CMB and try to see if there are small departures from *perfect* isotropy." Now what made people look for such anisotropy, though small? A very simple consideration. Astrophysicists asked: "Today's Universe is full of galaxies. There are literally billions of them. How did the galaxies form? How did the matter that was originally distributed uniformly, clump together to provide the seeds for galaxy formation?" In other words, astronomers had the feeling that although the early Universe might have been pretty uniform, judging by the uniformity and isotropy of the CMB signal, surely there must also have been tiny fluctuations causing in turn ripples in the CMB. These ripples must be the signatures of density and temperature fluctuations in the early Universe. And these fluctuations must have provided the original seed for galaxy formation.

16.4 The BOOMERANG experiment

Conjectures are nice but they have no value unless put to test. And so in the early nineties of the twentieth century, a satellite named COBE [COsmic Background Explorer] was launched in America. COBE's mission was to see if there are indeed tiny fluctuations in the CMB intensity and temperature over small regions. COBE scanned the sky carefully and showed that temperature fluctuations of around 30 microKelvin did exist over angular intervals of around seven degrees. This result was considered very significant because it showed that while CMB is indeed by and large uniform and isotropic, it *did* have tiny ripples indicating a certain amount of non-uniformity in the early Universe. This initial non-uniformity was believed to be enough to seed and trigger subsequently the clumping of matter, later to form into galaxies.

They say that the appetite grows by what it feeds on—the more one eats the more hungry one feels! This may not be quite true where the stomach is concerned, but where science is concerned, it is certainly true. Having tasted 'blood' with COBE, scientists now began to plan new satellites that would give much more detailed results about anisotropy. Two satellites are in the pipeline, firstly MAP (Microwave Anisotropy Probe) to be launched by America and PLANCK to be launched by European countries. MAP, for example, would be able to record temperature differences of 20 microKelvin and with an angular resolution of about 0.2 degrees across the sky (compared with the 7 degree resolution of COBE). The PLANCK satellite would, when it comes, have even better capabilities. Careful studies of angular anisotropy are important as they would lead to a good estimate for Ω.

Meanwhile an important experiment known as BOOMERANG has already pinned down Ω to a great extent. The word BOOMERANG is an acronym for Balloon Observations Of Millimetric Extragalactic RAdiatioN and Geomagnetics. The BOOMERANG experiment involved sending a balloon up to the stratosphere to a height of about 38 km. The balloon carried equipment to monitor CMB temperature in different directions. The balloon was launched in Antarctica and remained afloat for about ten-and-half days. The BOOMERANG map of CMB is 40 times more detailed than that of COBE, and corresponds to an angular resolution of roughly 0.25 degrees.

The BOOMERANG data was analysed the way angular distributions usually are. Here is a simplified account. Suppose one has a quantity $f(\theta)$ that varies as a function of angle θ. One knows that $f(\theta)$ can be expressed as

$$f(\theta) = \Sigma_l a_l P_l(\cos\theta).$$

The coefficients a_l contain the essence of the angular variation of the function $f(\theta)$. I am sure this expansion reminds you of Fourier expansion; in fact it is in the same spirit.

Figure 16.2 shows how the coefficients associated with the BOOMERANG data vary with l. Such a graph is referred to as the angular power spectrum. The results have been analysed using models with various values of Ω, and they strongly favour a value for Ω around 1 (we shall come to that presently). This sends a strong signal

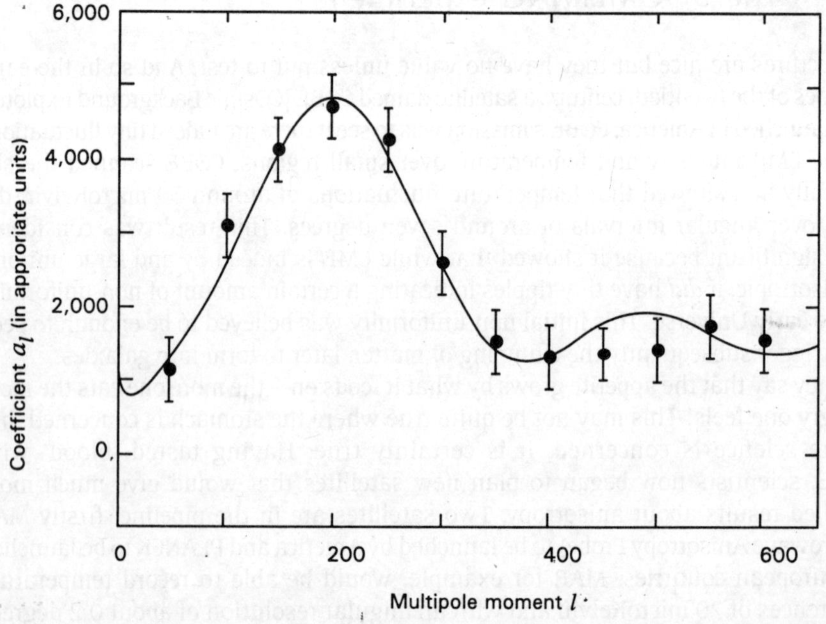

Fig. 16.2 *This figure shows the results of the BOOMERANG experiment. What is plotted is essentially the coefficient a_l, as a function of the multipole moment l. The solid curve is based on a model calculation with $\Omega = 1$. It is the best fit for the BOOMERANG data.*

that if matter including dark matter cannot make Ω equal to 1, then other agencies must pitch in. Incidentally, note that this data of BOOMERANG pertains to the early Universe; so, by implication we are trying to read the value of Ω for that time. If Ω varied with time, we would be in trouble!

16.5 What the supernova says

Recent data on supernova explosions also strongly support a value of Ω close to 1. I hope you recall what a supernova explosion is from ***Chandra***. Essentially, it is an exploding star. In recent years, a careful study of the redshift of supernovae has been made. Why study the redshift? Because the redshift gives an indication of what is happening to the expansion rate of the Universe. Theoretical studies show that the redshift varies with supernova distance as indicated in Fig. 16.3. Let us try to understand the significance of these curves with a simple analogy.

Let us suppose we pull a rubber band and take two snaps of it – see Fig. 16.4(a). The displacements of the different points on the band are marked with the

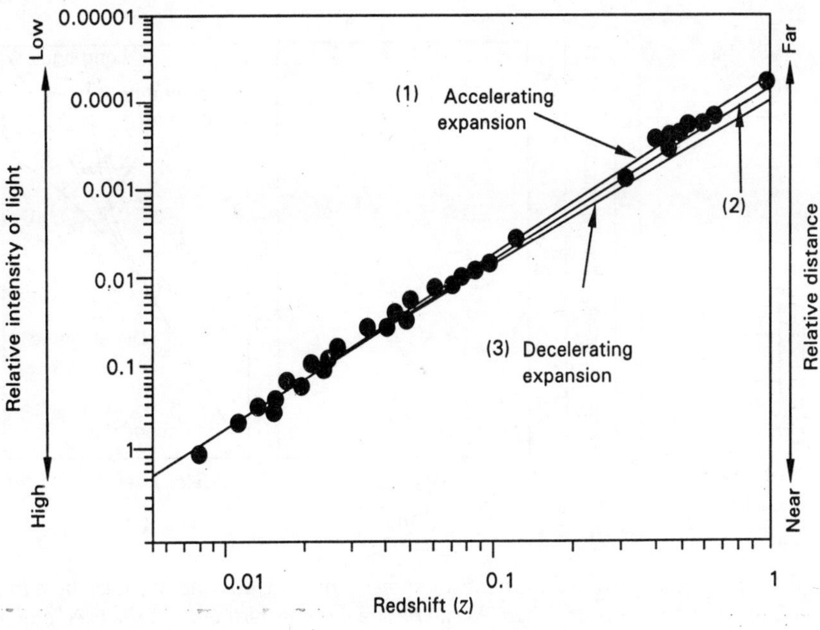

Fig. 16.3 *This figure shows schematically curves predicted for the redshift of distant supernova, assuming three different scenarios. The scenarios are: (1) long ago the Universe was expanding much slower than it is now, (2) the expansion rate has remained constant, and (3) that long ago the expansion rate was higher than now. Experimental data points obtained from a study of supernova explosions seem to cluster close to curves 1 and 2. Many astronomers seem to feel that between these two options, curve 2 is perhaps closer to reality.*

help of the arrows. This also tells us that the longer the arrow the greater is the velocity. From this we see that points close to the origin move with smaller velocities than the more distant points. With a knowledge of the velocities of the various points, we can draw a graph of velocity versus distance; such a graph would be a straight line provided the rate of pull is constant. For different constant pull-rates, we would get various straight lines as in Fig. 16.4 (b).

We now have to carefully carry over these ideas to the case of supernovae. In doing so we must keep in mind two facts. Supernovae are at long distances from the Earth. Therefore, the light that comes to us from them corresponds to light that was emitted long ago. If the expansion rate of the Universe has remained constant throughout, then there is no problem. But if the rate has varied, we have to be careful. Say the expansion rate was less in earlier times than at present. This means that the expansion is now accelerating. So, if we try to make a velocity profile for the galaxies, then for very distant galaxies we must use the curve marked slow in (b) and use the lower curve for nearby galaxies. In other words, when we use the schematic of (b) and take it over to the cosmos, we must adopt the curved line shown dashed in (b). In this way we can make up three curves as shown in Fig. 16.4(c), corresponding to an accelerating Universe, a Universe expanding at a constant rate, and a slowing-down Universe respectively. It is the set of curves in

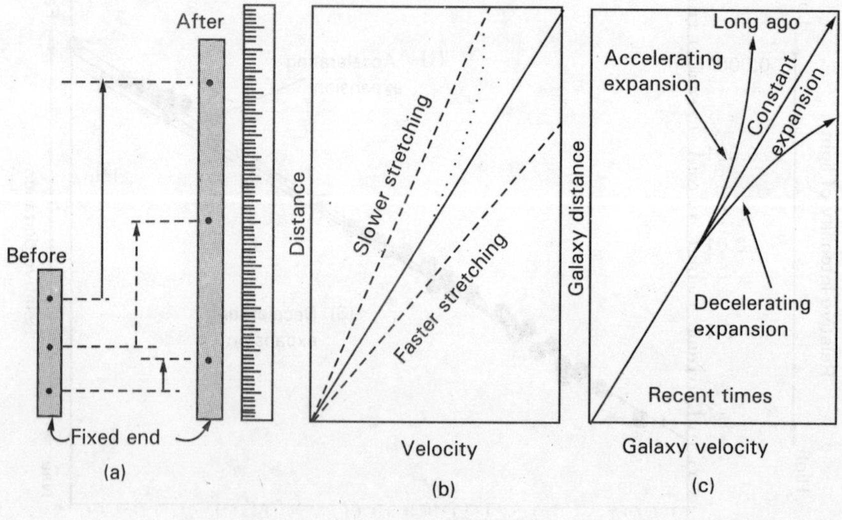

Fig. 16.4 *This figure provides an understanding of the previous one. We have here in (a), a rubber band that has been stretched. As can be seen, points farther from the fixed end move to a greater extent. Let us say that it takes one second to do the stretching. Then clearly, the velocity of movements of points on the band would depend on the location of the points. In (b), we have velocity vs. distance graphs for different (constant) rates of pulling. The results of (b) can be used to guide us about what to expect in the case of galaxies. For example, if the Universe expanded at a slower rate earlier than it is doing now, then we have to cross over as shown by the dotted line for reasons explained in the text. With this background, we can prepare a qualitative plot as in (c) that gives meaning to Fig. 16.3. Note that while (c) above is a linear plot, that in Fig. 16.3 is a log–log plot!*

(c) that provide an understanding of the data of Fig. 16.3 when we remember that for galaxies the velocity is synonymous with redshift.

Going back to Fig. 16.3, the experimental results seem to definitely rule out the curve marked 3. So it seems like a toss between the remaining two curves, that is either the expansion rate remains unchanged (curve 2) or it is accelerating (curve 1). Accelerating means that the *Universe is expanding faster today than it was before.* The odds for the former (namely the expansion rate has remained constant) seem to be slightly higher, though many would not rule the possibility that the Universe is in fact accelerating faster today than in the past.

16.6 Putting it all together

We now pause and take stock, putting the BOOMERANG and the supernovae data together. This is done in Fig. 16.5. This figure is really a map of space. The two axes show Ω_m, the contribution to Ω from matter, and Ω_Λ, the contribution from non-material source. Two comments: firstly Ω_m includes all types of matter, both luminous and dark. Second, Ω_Λ is nothing but a relative of our old friend the cosmological constant that Einstein first introduced in a hurry and later disowned totally. The exact relationship is

$$\Omega_\Lambda = \{\Lambda/3H_0^2\}.$$

For many decades Λ was considered a big joke, but now it has come back with a vengeance and with great respectability too. I wonder what Einstein would say about this strange resurrection of the CC that he had once dropped like a hot potato.

We shall now discuss the content of Fig. 16.5. Basically we see two bands from two independent experiments. The heartening thing is that there is a region where the two bands overlap. And in the region of overlap, $\Omega_m + \Omega_\Lambda$ is close to unity. The message is loud and clear. The overall value of Ω for the Universe is close to unity. In other words, the Universe is in all probability flat. Twenty years or so ago, there was much debate about this, and one even thought that perhaps the Universe may start contracting at one stage and end up in a Big Crunch (see ***A Hot Story***). But today, most cosmologists and astrophysicists favour a flat Universe.

We next consider yet another map that gives a slightly different perspective on the Ω story. Here we examine the contribution to Ω from matter (luminous and dark) and dark energy separately. This results in Fig. 16.6. Once again, there are bands and also a region of overlap that is pretty close to the line corresponding to a flat Universe.

Let us take this a bit further. What the CMB results indicate, and they do this pretty strongly, is that the geometry of the Universe is Euclidean; in other words, the Universe is flat. Remember, being flat does not mean that there is no expansion. There is expansion, and it is like stretching a flat rubber sheet uniformly in all directions (recall Fig. 6.3). But now let us ask: How come the Universe is flat

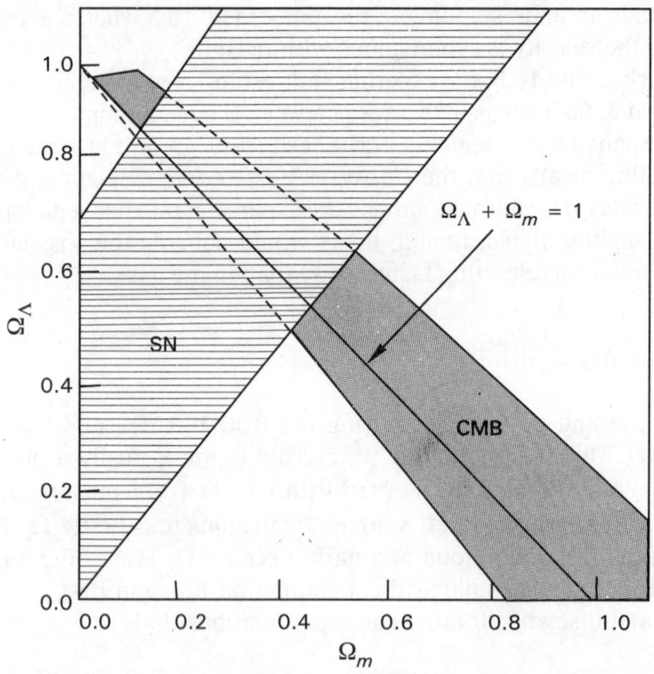

Fig. 16.5 *This is another perspective of the BOOMERANG results. What is shown here is the space bounded by Ω_Λ and Ω_m. The solid sloping line corresponds to $\Omega_\Lambda + \Omega_m = 1$. There are two shaded regions in the figure. One of these represents the region of this space that is consistent with supernova (SN) redshift data. The other represents model calculations based on the BOOMERANG (CMB) data. Notice that the two strips intersect over a region. What it means is that the actual values of Ω_Λ and Ω_m lie somewhere in the intersection region. The interesting point is that this overlap region embraces the solid line ($\Omega_\Lambda + \Omega_m = 1$). That is the reason why one now feels very confident that Ω is indeed close to 1.*

when matter and dark matter contribute so little to Ω? This, I pointed out earlier, is because there is an additional contribution to Ω coming from an unknown source of energy. Since energy is equivalent to mass, we can perhaps say that this unknown source of energy/mass lifts Ω to a value equal to 1.

16.7 Vacuum energy and quintessence

People have been thinking a lot about this new entrant Λ and what exactly it means. As I just now mentioned, Λ is now sought to be explained in terms of energy. You are aware that quantum field theory admits vacuum, and, as I have already explained elsewhere, the vacuum is not empty but full of virtual particles. The vacuum energy associated with these virtual particles, the experts then declared, was the source of the new avatar of Ω. This is really amazing because Ω is really a child of classical theory, i.e., general relativity, and now it is being given a quantum mechanical basis!

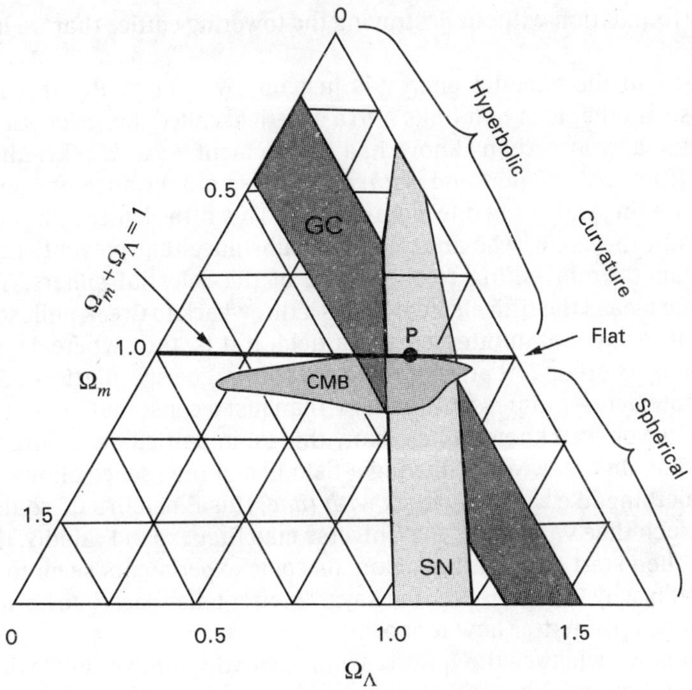

Fig. 16.6 *This figure offers yet another perspective on where we stand today with respect to Ω.. It is similar to the previous figure except that here, in addition to the Ω_Λ (dark energy) and Ω_m (matter, luminous and dark) axes, we have also the curvature of spacetime. There are three patches which show the regions consistent with galaxy clustering (GC) data, supernova (SN) data and the BOOMERANG (CMB) data. What is of interest is the region of common intersection. Notice that this region is close to Ω value of 1. For example, the coordinates of the point P are roughly $\Omega_\Lambda = 0.7$ and $\Omega_m = 0.3$, which is what we have also from Fig. 16.1.*

The idea that vacuum energy is the source of Λ sounds great but straightaway there was a huge problem. We need a value of about 0.7 for Ω_Λ but do you know what value the experts got for the vacuum contribution working with the Standard Model and all that? Only 10^{55}! There were other estimates that give a value as high as 10^{120}! No wonder one scientist has described this discrepancy between theory and experiment as a stupendous failure! However, says this scientist, this failure need not cause any embarrassment; it just means that there is some great physics lying beyond. His point is simply this.

The Standard Model works great in many areas. But when we try to apply it to get at the energy of the vacuum and thus an estimate for the dark energy as it is being called, we are off by an astronomical margin. This suggests that we are perhaps "dumping too much 'garbage' into the vacuum. The Standard Model which has a large number of fluctuating quantum fields including a Higgs field, is an egregious polluter of the vacuum. There is no doubt that the Standard Model has a highly beautiful and successful structure, but it may be based on a conception of vacuum or spacetime that is flawed. It is our challenge to repair that faulty foundation without destroying the towering edifice that we have built upon it."

Linking Λ to the vacuum energy is just one way of giving it a meaning. There is also another idea that links Λ to a new field called the *quintessence field*. Our ancients, as you perhaps know, had five elements: *akasa* (sky/ether), *vayu* (air), *agni* (fire), *jala* (water) and *prithvi* (earth); so did the ancient Greeks. They had earth, air, fire, water and one more element. This fifth element (the equivalent of our *akasa*), they said, was an ephemeral substance that prevented the Moon and the planets from falling to the centre of the celestial sphere. The word quintessence means the fifth element and is a throwback to Greek philosophy. It is all very well to say that quintessence is a field; if it is, then where does it come from? What is its origin? At present there are only vague speculations. One thing you must note here—quintessence is more than just a constant like Λ. It is a field and a field as you well know can do many things; in particular, its strength may vary with time. In other words, having a field like quintessence allows room for having something like a Λ that *varies with time*; this Einstein's CC could not do. Thanks to such time variations, the Universe may first expand rapidly, then slow down, and then start accelerating again, as some experiments seem to suggest. Perhaps that is why Professor Wheeler says: "Matter tells spacetime how to curve, and spacetime tells matter how to move!"

No one knows whether the quintessence idea will survive, but one thing is certain; the CC is no longer the disrespectable constant it was once considered to be by the very man who invented it!

16.8 Summary

- The future of the Universe is unquestionably connected with the value of Ω.
- Estimates of Ω based on luminous matter content of the Universe give a value much less than 1.
- Inclusion of the contribution of dark matter raises the value of Ω somewhat, but still does not bring it anywhere near 1.
- On the other hand, BOOMERANG and other experiments clearly point to a value ~ 1 for Ω.
- So, to raise the value of Ω, some have resurrected the cosmological constant Λ.
- Attempts have been made to interpret Λ in terms of the vacuum energy. But estimates of the vacuum energy based on the SM are astronomically high compared to the value needed. This shows that SM is way off where Cosmology is concerned.
- Attempts have also been made to interpret this additional contribution to Ω in terms of the quintessence field.
- It would be some time before a clear picture emerges about how exactly Ω achieves a value close to 1.

- On the other hand, astrophysicists who are studying supernovae are not yet totally sold out on a flat universe. They think we many even have a Universe that is expanding away to glory. Right now their data certainly permits that possibility, within the limits of experimental errors that is. So who knows, Ω may even be different from 1!

To repeat, there are essentially three quantities that are crucial for specifying uniquely the geometry and the evolution of the Universe. They are: The Hubble constant, the mass density of the Universe, and finally the age of the Universe. All that I have described here are based on *current* estimates for these crucial quantities. The story may yet change!

17

THE LION AND ITS TAIL

This chapter deals with what might be called the Second String Revolution when String Theory staged a comeback, leading on to more exciting theories wherein membranes replaced strings. Looks like the game is just beginning, and it promises to be very exciting!

Nature shows us only the tail of the lion. But I do not doubt that the lion belongs to it even though he cannot reveal himself because of his enormous size.

Albert Einstein

17.1 Introduction

This, for me, is a very difficult chapter because it is like talking about the entire lion, when all that one knows about is its tail. To put it differently, I have to tell you about so many different things that have happened in recent times, most of which I barely understand! Then why attempt this chapter at all? Just so that I can give you a glimpse, even if inadequate, of the tremendous excitement sweeping the field. As for the future of physics, to me it seems as if we all are assembled before Ali Baba's cave and that the magic formula for opening the cave is almost at hand. I could be totally wrong of course but as I survey the scene in amazement, that certainly is the feeling I get. Starting from the simple electron and basic radioactivity (which I used to think was exciting stuff when I was in college), we have come a long, long way in just one century. This chapter would, I hope, give you some idea of where we stand at present and also of the *tremendous* ground covered since J. J. Thomson detected the electron.

In Chapter 5, I presented an outline of how people moved from point particles to strings as the basic building blocks of matter in physics. To recap briefly, in String Theory one replaces the myriad of particles with a short string that can have many 'modes', like the usual vibrating string of classical physics. These modes correspond to the different particles. If one compares these different modes to different musical notes, then, as someone said, the Universe may be regarded as a 'symphony of vibrating strings'.

When it first appeared, String Theory attracted attention because there was in it a particle that had all the attributes of the graviton. What this meant was that in String Theory, gravity did not have to be put in by hand; rather, it was a *natural* component of the theory. As the high priest Witten once remarked, "Unlike conventional quantum field theory, String Theory *requires* gravity. I regard this fact as one of the greatest insights into science ever made." The next step was to integrate SUSY into String Theory, which then became Super String Theory. With this, one had room not only for bosons as in ver. 1.0, but also fermions. Super String Theory was thus ver. 2.0 of String Theory.

The advent of ver. 2.0 produced a lot of excitement, especially with some people hailing this new avatar of String Theory as the *Theory of Everything* (TOE). However, after a while interest cooled off somewhat for a variety of reasons. Basically, there were too many parameters in the theory and too many variations of the theory that looked equally good. The period roughly from 1981 to about 1984/1985 when there was a small boom, might be called the *First String Revolution*. After that a lull, and towards the mid-nineties there was once again a great spurt of activity in the String Theory area. This might be called the *Second String Revolution*, and the present chapter is mostly about that.

Now one of the reasons why many physicists lost interest in Super String Theory was that there were two principal unanswered questions: 1) Where do strings come from? 2) Is our Universe among the many solutions that String Theory throws up? It looks like that these questions may soon receive answers and that is what has produced a great deal of excitement. A recent development, the *M-Theory*, about which I shall tell you shortly, has got many physicists buzzing once again. As one physicist puts it, "I may be biased but I think it (the M-Theory) is perhaps the most important development not only in String Theory but in the whole of theoretical physics in the past two decades." Another physicist says, "In one dazzling stroke, M-Theory has come close to solving Super String Theory's two long-standing problems, leaving many theoretical physicists (myself included) gasping at its power!" So that's the kind of stuff we shall be discussing in this chapter, the promising new theory on the horizon!

17.2 A, B, C of Duality

It all started with the recognition of the important role of *duality* in String Theory. In figurative language, using the concept of duality is somewhat like looking at a coin from the other side. Mathematicians have for long have been used to the idea of dual spaces, dual lattices and stuff like that. For example, in solid state physics, we often go from the direct lattice (that describes the crystal structure) to the reciprocal lattice, which is very convenient for describing the phenomenon of diffraction of X-rays, electrons, and neutrons. In String Theory, duality has a deeper and more profound role to play. In particular, it gives information about the behaviour of String Theory in the strong-coupling regime. I shall say more about this soon.

Let us now try to understand what duality is all about, and why it is playing such an important role in string physics. We will begin with electromagnetic fields because that's where this business of duality really started. I have already told you in Chapter 10 that if magnetic monopoles are allowed for, then Maxwell's theory becomes symmetric if we make the set of dual transformations

$$\mathbf{E} \rightarrow \mathbf{B}, \quad \mathbf{B} \rightarrow -\mathbf{E},$$

$$q \rightarrow p, \quad p \rightarrow -q,$$

$$\mathbf{j} \rightarrow \mathbf{k}, \quad \mathbf{k} \rightarrow -\mathbf{j}.$$

Cut now to 1976 when Bogomol'nyi in Russia (then USSR) showed that the mass of all stable particles in certain class of gauge theories would be given by a universal formula of the form

$$M = \text{constant} \sqrt{(q^2 + p^2)}.$$

In the following year, Montonen and Olive in CERN showed that this places q and p on the same footing in an important way. Based on this, Montonen and Olive then made a conjecture. They said (in effect): "Let us say you have a theory for these particles. The electric charge q is a multiple of the basic electric charge e and likewise the magnetic charge p is a multiple of the basic magnetic charge g. There is an intrinsic symmetry between the electric and magnetic charges. Let us say in this theory, e is a parameter denoting the coupling constant. If e is small we say we have weak coupling and we know how to get results. We basically use a perturbation theory expansion in e (I have discussed this sort thing before in ***QR II*** and also in Volume 1). However, when e becomes large we are in trouble because we are now in the strong coupling or the non-perturbative regime, i.e., a regime where perturbation theory fails. But this precisely is where we exploit the symmetry noted earlier. Thanks to that symmetry, e and g have a reciprocal relationship, meaning

$$e \sim (1/g).$$

What this says is that when e is large, g is small and vice versa. Since e and g are on equal footing, we might as well solve the problem using g as the coupling constant, instead of struggling to do a horrendous, non-perturbative calculation involving a large e. In other words, *the duality inherent in the symmetry of q and p allows us freedom to navigate between using either* e *or* g *as the coupling parameter, as convenient.* If e is large, we switch to problem-solving using g since g is small (which means we can do a perturbation expansion in g). On the other hand, if g is large, we start using e as the coupling parameter."

This is an important conjecture. Earlier, people just noted the symmetry of the extended Maxwell's equations, the presence of the monopole and things like that. But Montonen and Olive went further boldly. Let me now explain the implications of what they suggested. Normally we take the electric charge to be fundamental and treat it as a point particle. It has point interactions and when we want to explore different domains of interactions, we treat the basic electric charge as a parameter related to the coupling constant. The formal coupling constant is the fine structure constant α_e [$= e^2/(4\pi c)$] that must be familiar to you from earlier volumes. In this picture, the magnetic charge is a *composite particle with a core*. Further, when the electric charges interact weakly, the magnetic charges couple strongly. If α_m denotes the magnetic coupling constant, there is the reciprocal relation

$$\alpha_e \sim (1/\alpha_m).$$

From this it should be clear that when the electric coupling constant is weak the magnetic coupling constant would be strong, and vice versa.

OK, suppose all this is true; then what? Well, there is a big bonus if the conjecture is true. Say, one is trying to do calculations in a regime where α_e is very large. We are stuck and we don't know how to proceed in this strong-coupling regime. This is where duality provides an escape route. Forget α_e and work instead with α_m. Working with α_m is no sweat at all since it is small, and we are in the weak-coupling regime where α_m is concerned. Physics being the same—and that is the essence of the duality conjecture—what does it matter how we do the calculation? Why bother about α_e when α_m is available? That is the flexibility that duality offered. In simple terms, all that I have said above could be paraphrased in the following crisp statement: *The weak coupling for one picture is the strong coupling for the other and vice versa.*

We must in passing also note that the physical picture also gets somewhat changed when we invoke duality. Here is an example. Let us say that point electric charges are basic. When we have these charges moving around, we have a magnetic field. We can also think of these magnetic fields as originating from magnetic monopoles. What is the relationship between the moving charges we had and the monopole? The so-called monopole is nothing but a complex cluster of moving charges. Suppose we knew nothing about electric charges and knew only about monopoles. We could then start assuming that monopoles are point particles. When many of them tangle and move about, they generate electric fields. We could now say, "Well, why can't I think of these electric fields as coming from electric charges?" Sure one can have a picture like that. Next question: "What do these charges look like?" Answer: "They are complex clusters of monopoles!" This is the way duality works.

Are there problems where this kind of facility would be particularly useful? Sure. Let us take the quark problem. We know that when quarks are close to each other they interact weakly. The coupling constant is small and perturbation theory can be used legitimately. But when quarks are far apart (and this is the kind of situation we face when we try to tear two bound quarks apart), the force between the two becomes large (see Chapter 11, Volume 1). In this case, one cannot use perturbation theory. People in the old days used to just throw up their hands in despair not knowing what to do. This precisely is where duality could come in handy. Why was the quark problem difficult at large distances? Because the coupling was strong. Duality offers a nice escape route. It enables one to transform the original problem to another problem and in this new version the coupling is small! Since the coupling is small and weak, the calculation is in principle simple, and in this way an answer could be found. The beauty of all this is that the physics does *not* change when one adopts the dual view—it is like saying that the physics of the crystal is the same whether it is described in terms of the direct lattice or the reciprocal lattice.

In essence what Montonen and Olive suggested was that where QCD is concerned, either the colour charge or the colour monopole approach to the theory should give the same results. As such, from a practical point of view' it is desirable to choose that approach which is easy – see Fig. 17.1. In practical terms, this means that if one can find the dual theory to QCD, then one could perhaps explain many aspects of high-energy behaviour of mesons and baryons using this dual theory.

Here I would like you to note one very important point. You might ask: "The colour monopole is really a composite object. How come one is, in the dual view, able to regard it as an elementary object and get away with it?" The answer is the following. True, from the point of colour charges the colour monopole is indeed a composite object. However, and this is the important point: from a *dual symmetry* point of view, the colour monopole is in fact an elementary object. In other words, *duality transforms geometric complexity* (if I may put it that way) *into a kind of higher-order simplicity.* Thus, when convenient, use quark colour charges as the basic object and work with them. In this picture the quark monopoles are composite objects. If the problem gets tough because quark charge-coupling is strong, switch over and use quark monopoles as the elementary objects. The original troublesome strong-coupling constant now gets replaced by another one that is weak, and one should be able to sail through! (Of course, the quark charges now become composites of elementary quark monopoles.) That is roughly about it. Got it?

OK, all this is an idea. What about practice? Well here the going gets tough because there is the question: "How do we *know for sure* that the Montonen–Olive conjecture *actually* works?" We can say it works if, for a particular problem with coupling constant say α, we have the correct answer not only in the weak-coupling but *also* in the strong-coupling limit. We could then do *another* calculation using $\alpha' \sim (1/\alpha)$ as the coupling constant (remember this is the dual coupling constant) in the weak-coupling limit and then compare the two results. If duality holds, then the answer we get with α large and using non-perturbative methods must be the same as the answer we get with α' and the perturbation method. This check could not be done for any real problem, and so the Montonen–Olive idea remained just that, a mere idea, for some time at least. In fact the idea was greeted with much scepticism.

17.3 Duality takes firm root

Montonen and Olive gave their idea a long time ago, way back in 1977. Later when people started taking superymmetry rather seriously, it became clear that

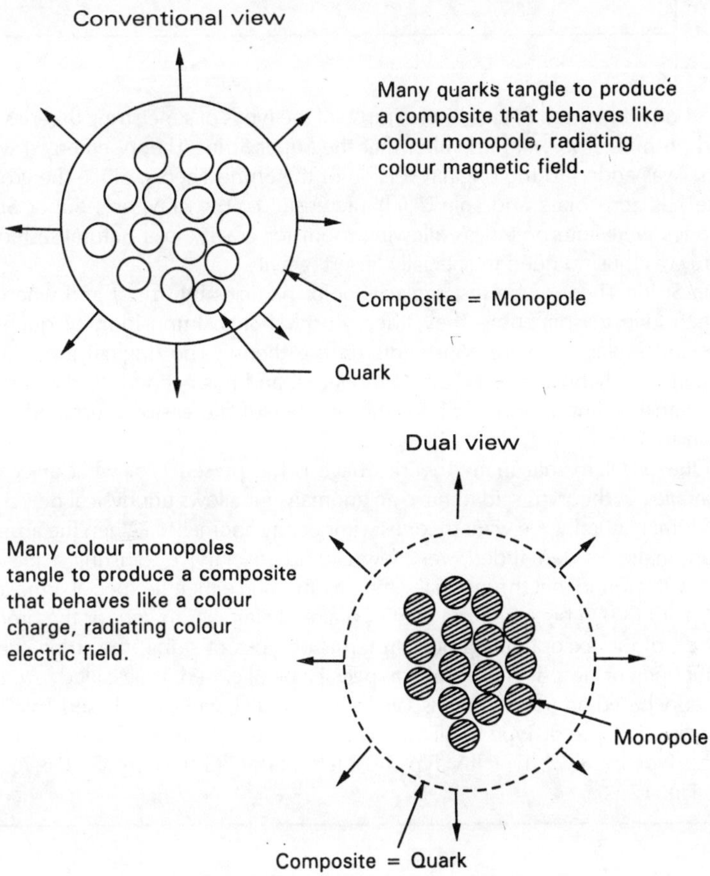

Fig.17.1 *This figure illustrates how duality provides two complimentary points of view of the same phenomenon. When the quark moves it produces a colour 'magnetic' field. Sometimes, many moving quarks can tangle into a composite object that has a colour 'magnetic' monopole moment. This is one way of looking at things in which the quark colour charge is primary and the quark colour monopole is derived. Alternatively, we can start as if the quark colour monopole is primary and regard the quark colour charge as a composite object that is formed by many monopoles tangling together.*

the Montonen–Olive duality was more plausible in this case. So what happened after that, especially in Super String Theory, that gave a kick-start to the Second String Revolution? To get a feel for that, let us pause and look at the 'scoreboard' of String Theory as it stood around 1985–1986. Firstly, there were five strong candidate string theories, in ten dimensions that is (see BOX 17.1). Which was the 'real McCoy?' No one knew. Each theory seemed good in its own way. Next, people started ompactifying the extra dimensions. The theory has 10 dimensions but we need only four (3+1), i.e., three space and one time dimension. Reducing 10 to 4 means that 6 dimensions have to be compactified. When this was attempted it turned out there were literally millions of four-dimensional String Theories! With one starting theory leading to a million final theories, it was as if the cure

BOX 17.1

This box is intended to give brief glimpse of the types of superstring theories that hold promise. You will remember that the Super String Theory emerged when SUSY was added to the original version of the String Theory. Since the ground state has zero mass and spin 2 (characteristic of the graviton), Super String theories were interpreted as allowing room for gravity in a natural fashion—there was thus no need to specially 'insert' gravity.

In String Theory, one first replaces point particles by strings and when the latter move in spacetime they trace world sheets. Imposition of quantum mechanics places severe constraints on the theory. The original formulation permitted only bosons to be accommodated, and this needed 26 dimensions. But with the imposition of SUSY, the number of dimensions required came down to 10.

One problem that many theories have is the presence of what are called *anomalies*. A theory is said to have an anomaly if it allows unphysical degrees of freedom. Naturally, we want theories without any anomalies. When the absence of anomalies is demanded, very few candidates survive. Slurring a bit over details, it is found that the internal symmetry must be either the special orthogonal group *SO* (32), a bigger cousin of *SO* (3) already familiar to you, or the product $E_8 \times E_8$ of the exceptional group. In terms of types of string, the string can be either open or closed. There is also a special type of closed string called *heterotic*. A theory based on open strings is said to be of Type I, while one based on closed string is said to be of Type II. All things taken together, one has the following five types: Type I *SO*(32), Type IIA, Type IIB, Heterotic *SO* (32), Heterotic $E_8 \times E_8$. See also Fig. 17.2.

was making the disease worse! As someone put it, it was an embarrassment of riches.

It was at this juncture that Asoke Sen, then at TIFR, wrote in February 1994 a seminal paper in which he pointed that duality, when extended to strings (as opposed to point particles), helps not only in connecting different string theories but also in reducing the number of options. Suddenly the millions of theories did not seem like nonsense or a waste of time. They were trying to tell something and Sen's paper pointed out how to 'hear' them! In a sense this was the start of the turning of the tide. In March 1995, there was a conference in America on strings. Giving the opening talk, Witten, often hailed as the Pope of String Theory, used duality to draw together a large number of seemingly diverse ideas. The talk produced an electrifying effect, and the Second Revolution was now on.

In short, the string theory duality was not merely a route to escape from the log-jam of non-perturbative theories but a great unifier. To start with, it no doubt opened the door to problems that were previously considered intractable. But duality in String Theory did not stop with just this. It has brought about a total change not only in the way String Theory is to be perceived, but in fact in our approach to quantum mechanics as well, and that is what we have to absorb carefully. As one physicist put it, "String Theory might teach us a thing or two about quantum mechanics itself!" More about that soon. Meanwhile, let me now introduce you to the cousins of strings!

17.4 From strings to membranes

Physicists are sometimes so bold that they even give the impression of being rash (and also brash)! In 1986, tired perhaps of strings, Michael Duff decided to think in terms of a bubble that floated in a 11-D space, instead of working with strings that wiggled in 10-D space. This started a new fashion. Duff essentially said: "Why only strings? Why not membranes? Two-dimensional membranes? Why only two-dimensional membranes? There could be membranes in higher dimensions!" This slowly started a new fashion—membranes instead of just strings. These multi-dimensional membranes are simply called *branes*. In this language, a point particle is a zero-dimensional brane and a string is a one-dimensional brane. From strings, people had now graduated to higher-dimensional branes. One special membrane 'surface' is of the type discussed long ago by the mathematician Dirichlet. Membranes with such a surface are referred to as *D-branes*. Of course, for a while the brane idea seemed like a just a gimmick to many but Witten's talk seems to have changed that view. Thanks to Witten, branes now became the in-thing! And as a result, the Theory of Strings has given place to the *Membrane Theory* or *M-Theory* for short.

The floodgates having opened, people have been busy discovering all sorts of dualities between all sorts of versions of string theories. At incredible pace, physicists have mapped almost all solutions and dualities that exist in 10, 8, and 6 dimensions (see also BOX 17.2). This is fine, but what we really require is such an exercise for 4 (= 3+1) dimensions, and that has not yet been done as far as I know.

One interesting point. The M-Theory specifically requires 11 dimensions whereas all the candidate string theories require only ten. Wasn't this an embarrassment?

Box 17.2

Here is an amplification of how duality helps in linking different solutions in different dimensions. The basic idea is the following: We start with theories in higher dimensions, and then come down to lower dimensions through compactification. And we want to see inter-relationships in this lower dimension —that is the game. Let us start with two string theories, A and B, in 10 dimensions. They are known to be dual. Now compactify A to say 6 dimensions; let us say that as a result we get theory A′. Similarly, we compactify B to 6 dimensions and this gives us B′. Thanks to M-Theory, we now know that A′ and B′ are dual of each other. What it means is that when A′ is difficult to handle, that is to say when we encounter the non-perturbative behaviour of A′, the problem can be tackled by dealing with B′ and vice versa. This is the way M-theory has been used to solve the entire class of problems previously thought to be unsolvable. This game is yet to be played out in 3 + 1 dimensions; that is what all are waiting for.

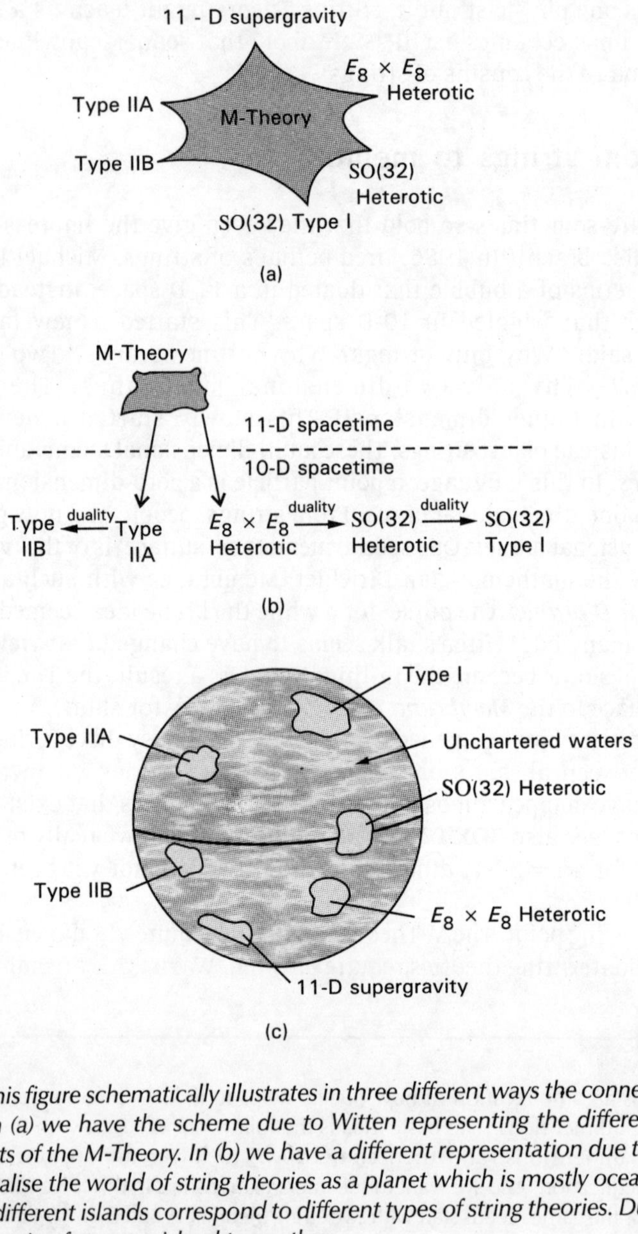

Fig. 17.2 *This figure schematically illustrates in three different ways the connectivity offered by duality. In (a) we have the scheme due to Witten representing the different theories as different limits of the M-Theory. In (b) we have a different representation due to Duff. Finally in (c) we visualise the world of string theories as a planet which is mostly ocean with several islands. The different islands correspond to different types of string theories. Duality is a boat that helps in going from one island to another.*

No! The Pundits simply said, "Well, you see, the eleventh dimension is compactified and so you really have only ten dimensions." They do magic like that! Perhaps that is why Witten says, "M stands for Magic, Mystery or Membrane, according to taste." Someone else who has gone gaga about M-Theory says that M stands for the Mother of all theories! One thing that the M-Theory approach has done is to 'tie' various types of string theories into one family, via dualities. This is

BOX 17.3

This is what Michio Kaku of City University of New York says: "Imagine three blind men hot on the trail of Einstein's lion. Hearing it race by, they give it a chase and desperately grab it. Hanging on to the tail for dear life, one feels its one-dimensional form and loudly proclaims, 'It is a string. The lion is a string!' The second man grabs the lion's ear. Feeling a two-dimensional surface, he calls out, 'No, no, the lion is really a two-brane.' The third bind man hanging on to the lion's leg, senses a three-dimensional solid and shouts, 'You are both wrong. The lion is a three-brane!' They are all right. Just as the tail, ear and the leg are different parts of the same lion, the string and various *p*-branes appear to be different limits of M-Theory. Townsend calls it *p*-democracy!"

schematically shown in Fig. 17.2. Basically, one imagines elementary particle physics to be like a world which is almost all sea with a few islands here and there. The different string theories are like the islands, and string duality is the boat that takes one from one island to another! See also BOX 17.3.

One day when the solutions pertaining to the 4-D world are properly mapped and examined thoroughly via duality, we would perhaps find that one of these solutions describes the known Universe and contains the Standard Model as a subset. Meanwhile, one physicist conjectures that this solution will 'descend' not from 11 dimensions (as M-Theory advocates would have it) but from a 12-dimensional theory which has been called the *F-Theory* (F for father!). Another physicist comments that there is an advantage in starting from higher dimensions for finding the correct theory for four dimensions. He compares this to what the Romans did during battles —they would seize a hill and survey from there. This hilltop was their 'hyperspace', from where they could see the battlefield as one unified whole.

So will the final theory be in 10, 11 or 12 dimensions? No one can say. The final theory may even turn out to be very different from what people are imagining at present. As one physicist put it, the present situation is rather like what one had in the old quantum theory, before the birth of regular quantum mechanics, I mean incomplete knowledge and some 'working rules' like the correspondence principle (see ***QR I***). I think this is a fair assessment.

17.5 Some exciting possibilities

The M-Theory approach has opened up a tantalising possibility. To understand it, first see Fig. 17.3. This shows that whereas the coupling constants of the three forces (weak, electromagnetic and strong) converge at an energy of about 10^{16} GeV, gravity acquires this value only around the Planck energy of 10^{19} GeV. There is thus a huge energy gap between the energy where the three forces 'meet' and the energy where the strength of gravity becomes comparable to the other three. This is conventional wisdom. However, using M-Theory, extra dimensions suitably compacitified and all that, it appears that the scale of Planck energy can be brought down to 10^{16} GeV. What does it mean? It means that thanks to the M-Theory the Planck energy gets reduced a thousand times, and

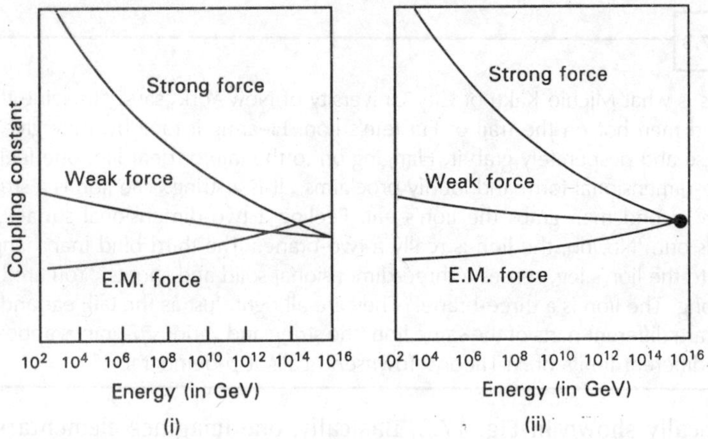

(a)

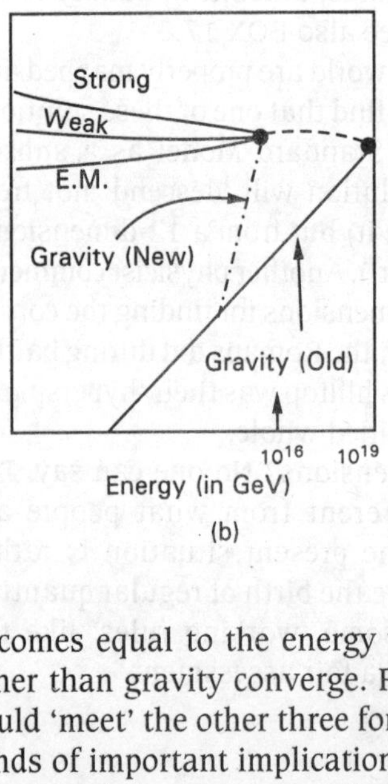

(b)

Fig. 17.3 *Study this figure carefully. It has been schematically presented earlier (in Chapter 2) but now is the time to take it seriously. In (a (i)) are shown how the coupling constants associated with the electromagnetic, the weak and the strong forces change with energy when there is no SUSY. They come reasonably close to each other at a high energy but do not actually intersect. With SUSY included, there is convergence of the three coupling constants at about 10^{16} GeV, as shown in (a (ii)). The coupling due to gravity which starts off rather weakly, picks up and catches up with the other three at about 10^{19} GeV. The new idea is that gravity meets the other three at the much lower energy of 10^{16} GeV. These ideas are sketched in (b). If this turns out to be true, it would most certainly have an important impact on physics.*

becomes equal to the energy where the coupling constants of the three forces other than gravity converge. Planck energy getting reduced means that gravity could 'meet' the other three forces at a lower energy. Obviously, this can have all kinds of important implications for Cosmology.

Meanwhile, a version of the M-Theory has been developed in which coordinates are represented not as simple numbers but as matrices. This matrix representation is reminiscent of the way matrices entered quantum mechanics. One may even say this is the new avatar of quantum mechanics suitable for dealing with gravity. It makes spacetime fuzzy, and this has profound implications.

Till we come to the physics of the very small, gravity and the other forces do not even greet each other. I mean when we discuss atomic, nuclear or even particle physics, we never consider gravity. But when we come to physics at very small distances, gravity enters the picture in a very significant way. And having entered, there is a sort of 'confrontation' between gravity and quantum mechanics. In earlier times the black hole, considered to be some kind of a representative of classical gravity and spacetime, used to challenge quantum mechanics. But now we are now reaching a stage where quantum mechanics is actually altering the nature of spacetime itself and dictating terms to gravity! Truly, physics is in the melting pot so to speak. What exactly is going to happen only the future can tell. All we can say now is that a tremendous shake-up is going on.

17.6 Enter black branes

A bit more about the shake-up just referred to. I mentioned earlier that this M-Theory business is providing deep insights into the interplay between gravity and quantum mechanics. An excellent example of this relates to what are called *mini black holes*. You will recall that Hawking argued that black holes, the authentic representatives of gravity, possess a property akin to entropy, and therefore radiate. Since then, using M-Theory ideas, people have been able to actually count the quantum states of a black hole formed by the collapse of an object originally made up of strings or branes (see BOX 17.4). For this reason, the branes that give rise to black holes are called *black branes*. From such investigations, M-Theorists could calculate the entropy, which agrees perfectly with Hawking's prediction. This is considered by many as a feather in the cap of M-Theory.

Not merely this. Black branes promise to solve the proliferation problem. Remember I told you that when the extra dimensions in string theory are compactified, one ends up with millions of 4-D string theories? Black brane advocates tell us that this problem gets solved thanks to a new topology—one might call it *quantum topology*—which is able to connect spaces with different number of holes, something unthinkable for mathematicians (see Box 17.5). What does this mean in physical terms? It means that the millions of 4-D string theories correspond actually to spacetimes with different numbers of holes. Imagine that! Spacetime now begins to look like a type of cheese that is full of holes, and is not continuous! OK, so we have these millions of different cheese pieces, each claiming to be a descendent of the original 11-D brane – so it seems. In reality, however, these cheese pieces are not exactly different; thanks to quantum topology, *they are mathematically equivalent;* only, they correspond to different energy states. So what happens now? This is how Michael Duff who was the first to go beyond strings to a bubble, sums it up:

> If all the spacetimes are thus related, finding the right one (i.e., the one related to the Universe in which we live) becomes a more tractable problem. The string may ultimately choose the spacetime with, say, the lowest energy and inhabit it. Its undulations would then give rise to the elementary particles and forces as we know them, i.e., the real world.

BOX 17.4

You will remember that the first phase of the quantum mechanical study of black holes started with the suggestion by Bekenstein that black holes must have an entropy S given by a relation of the type:

$$S \propto A/G\hbar,$$

where A is the area of the event horizon of the black hole and G is Newton's gravitational constant. Hawking was able to go further and demonstrate that black holes must actually radiate. This implied a specific value for the temperature T of the black hole, and when this result for the temperature was combined with the earlier result of Bekenstein, Hawking obtained

$$S = A/4G\hbar.$$

Basically, what Bekenstein and Hawking did was to give the thermodynamics of black holes. Now you must be aware from ***A Hot Story*** that thermodynamics has a microscopic basis namely, statistical mechanics. It is statistical mechanics that allows one to calculate, for example, entropy from first principles. Thus when we go to statistical mechanics , we find that entropy is related to the different ways in which a particular state can be realised.

The question arises: Is it in any way possible to calculate the entropy of a black hole by enumerating its states and counting them properly? In the mid nineties it was realised that at least mini black holes can be regarded as a collection of D-branes. Using then the machinery of duality, branes and all that, it was possible to do a quantum count and determine the entropy the way it ought to be. While the last word on the representation of black holes in terms of D-branes has not yet been said, there has certainly been a significant advance in the sense that the black hole, once regarded as the protégé of gravity, can now claim String Theory as its parent, at least where mini black holes are concerned.

BOX 17.5

Mathematics has become an indispensable part of physics and, by and large, the mathematics needed to describe some particular physical process or phenomena is almost always readily available, thanks to earlier work by mathematicians. A famous example relates to Einstein's theory of General Relativity.

Einstein realised that on account of gravity spacetime was curved; he needed suitable mathematics to describe that. Fortunately for him the mathematics of curved spaces had already been developed. Grossman helped Einstein in using this tool. But there also have been times, though few, when mathematics was not ready; and on such occasions, physicists (possibly because of impatience?) have resorted to self-help. A famous example is that of Newton who developed calculus, without which the study of dynamics would be so very difficult. Right now there is another such period when mathematicians, who generally tend to

Contd...

BOX 17.5 ...Contd

be rather patronising where theoretical physicists are concerned, are almost becoming jealous of what physicists are doing in their backyard! As you might expect, it is all connected with developments in String Theory, duality, and all that.

I remember way back in the mid-eighties of the twentieth century when the first revolution in String Theory was in progress, there was a lot of talk about the need for new mathematics. Not only physicists but even some mathematicians seemed to appreciate the need to open up new frontiers. At that time the theoretical physicists and mathematicians in TIFR decided to get together to have a serious dialogue. I was very happy when I heard about this because TIFR has always boasted of very good particle physicists and mathematicians. (I don't mean to suggest that other scientists in TIFR were sub-standard! Far from it; then and now, TIFR is a jewel where science is concerned! Just that at present my focus is on matters relating to String Theory.) OK, they had this intense dialogue; what happened next? Standard difficulty! There was the problem waiting to be solved, and there were these experts who, working together, could have done something about finding solutions. Only, they just couldn't come together. They worked in the same Institute, they occupied rooms on the same floor, in fact they went to the canteen together for lunch and tea, but when it came to professional work, there was this huge cultural divide. Mathematicians are very fussy and cannot tolerate what they regard as mathematical sloppiness. Physicists on the other hand, are eager and impatient to make Nature reveal her secrets. Impelled by the spirit of pioneers, they do not hesitate to rely on intuition and cut corners where rigour is concerned. A mathematician can never allow himself to give up rigour; to do so is like committing suicide! And so, thanks to this cultural divide, the get-together broke up, the mathematicians and physicists agreeing that their contacts would go back mainly to the social level as before!

By the way, this 'obsession' with rigour is nothing new. I have heard people say that when Dirac proposed the famous Dirac delta function (that physicists and even engineers use all the time), the mathematics community felt scandalised!

Let us move on to the Second String Revolution as it is called, concerned with duality and all that. This time, many talented physicists decided to take the bull by the horns, so to speak. "If the required mathematics is not there, let us develop it," they said. And they have done it too, done great things in fact. So much so that one of them, Edward Witten, sometimes hailed as the Pope of String Theory, actually ended up winning the prestigious Fields Medal. This medal is awarded once in four years to a mathematician for outstanding achievements, and is considered the Nobel Prize of Mathematics. Imagine that! A physicist, winning the 'Nobel' of mathematics! Naturally, mathematicians are stunned.

Now, how come a physicist was able to penetrate the privileged territory of mathematicians? He was able to, thanks to physics, yes physics! Physics gave him a quantum vision and helped him to dare. Mathematicians always think of space as smooth and continuous. So do physicists except when they come down to distances equal to the Planck length. Space now becomes 'spongy'

Contd...

Box 17.5 ...Contd

and all sorts of things begin to happen which are forbidden in continuous spaces.

Here is a stunning example. I presume you have heard of topology, a branch of mathematics that is sometimes called rubber-sheet geometry. In simple terms, the spirit of topology can be understood as follows. On a plain paper, a circle is different from a square; there is no way a circle can be converted into a square and vice versa. Now suppose that a square is drawn on a rubber sheet. By stretching the sheet suitably in various directions, one can in principle convert a square into a circle. Thus, for a topologist, a circle and a square are not different from each other, though for a geometer they are. Similarly, suppose one considers a cube made of putty, a pliable material. It is evident that by pressing and squeezing suitably, the cube can be transformed into a sphere. Thus the sphere and the cube are equivalent in topology. On the other hand, there is no way a sphere can be transformed into a doughnut and vice versa; thus, a sphere and a doughnut are supposed to be topologically non-equivalent – this is conventional mathematical wisdom. And do you know what? If space is permitted quantum effects, then a doughnut can be thinned down at a point, pinched, and then remoulded into a sphere – see figure below. What was a no-no for mathematicians has become a yes-yes for physicists, after endowing space with quantum properties.

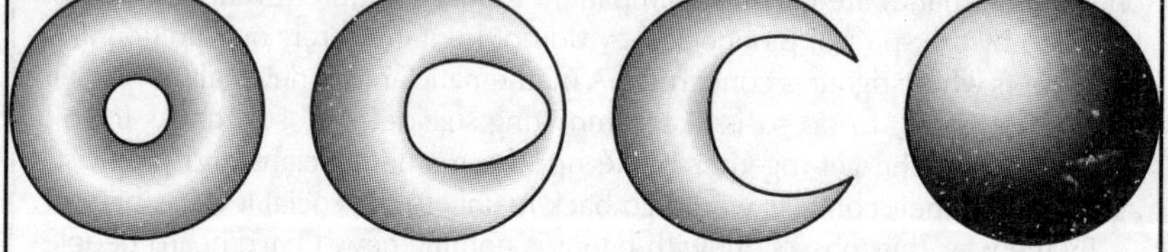

Mathematicians are no doubt stunned. Grudgingly, they concede that these are new results that they had never thought of; yet they are suspicious and will accept them only if they can prove String Theory results to their satisfaction, that is, with the rigour that testifies to mathematical purity. They say, 'Our entire culture is based on rigour and how can we give that up?' But physicists who are out to extract the secrets of Nature often go more by intuition and are free from the burden of rigour.

This reminds me of an incident in a cricket coaching camp held in Madras about fifty years ago. The coach was from England and he was very fussy about footwork in batting. One day at the nets, a batsman hammered a ball right out of the ground and was feeling very pleased with himself. The coach, on the other hand, threw his hands up in the air in exasperation and shouted, "Oh no, your footwork is all wrong!" The batsman was annoyed—naturally—and shot back, "Damn the foot! Where is the ball?" Looks like that here, does it not?!

17.7 A new version of Newton's law of gravity?

The story does not end here. Investigation of branes and the way gravity operates in a world that is produced by compactification has led to yet another strange and

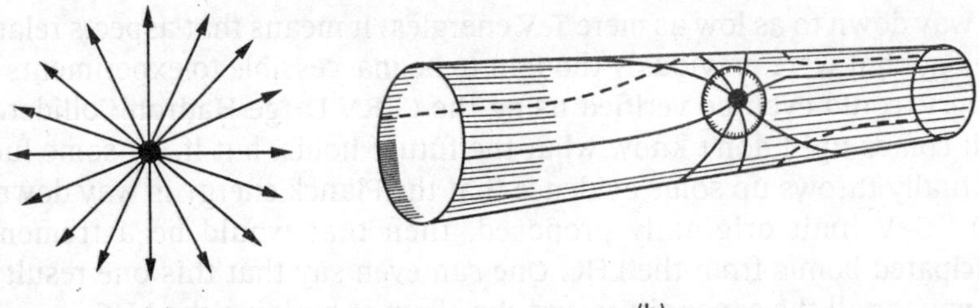

(a) (b)

Fig. 17.4 *This figure illustrates an exotic new concept namely, how gravity could possibly have different behaviours at different length scales. (a) shows how the gravitational lines of force radiate from a mass in accordance with Newton's law. This is conventional behaviour. (b) shows something more complicated, though in a very schematic manner. Here, the small circle is representative of the space we live in, while the cylinder shows the totality of space, i.e., the ordinary space and the extra dimensions. There is a mass at the centre of the circle and lines of force go out as indicated. Within the circle they appear to diverge radially; so it seems as if Newton's law holds. But gravity spreads to the extra dimension also. When it does, the lines of force eventually end up travelling parallel to the axis of the cylinder—a very different behaviour compared to that expected from Newton's law.*

interesting idea. To understand this, you must refer to Fig. 17.4. Figure 17.4(a) is a highly schematic drawing of how lines of force due to gravity spread out from an object with mass. If one looks at a *small* region around the mass in a multi-dimensional space, the lines seem to go outward radially. This precisely is what one expects from Newton's law – see Fig. 17.4(a). Suppose one looks at what happens at *larger* distances. The lines now start wrapping around the cylinder, and move along parallel lines – see Fig. 17.4(b). Clearly, this pattern is not representative of the familiar inverse square law, and all this happens on account of the overall nature of spacetime. Here, over a small region spacetime seems to be flat, but over a larger region it is cylindrical.

Scientists say this tells us that in some models we may have the following strange situation: While in 3-D the inverse square law is what we expect of gravity, the law would be different when there are many extra dimensions. Let us say that theory has n extra spatial dimensions, that is, there are in all (n + 3) spatial dimensions. The n extra dimensions then compactify into small circles of radius R. In such a situation, instead of the familiar ($1/r^2$) form, at short distances the force of gravity would now go like

$$1/r^{(2+n)}.$$

That is, at short distances in 3-D, gravity would follow an inverse $(2 + n)$ law. This is something one had never heard of before. People believe that R could be about 1 mm or less, and experiments are under way to check whether this new law actually holds. No one has so far tested Newton's law at such short distances, and our belief that it hold at all distances is merely a conjecture extrapolated from the observed behaviour at relatively large distances.

There is more. If all the above is true, it appears that the Planck energy would be brought way down not merely from 10^{19} GeV to 10^{16} GeV as hinted earlier, but all the way down to as low as mere TeV energies! It means that aspects relating to Cosmology that were previously thought to be inaccessible to experiments in the laboratory, could even be verified using the CERN Large Hadron Collider (LHC) when it comes up! I don't know what the future holds, but if, by some luck the LHC actually throws up some evidence that the Planck energy is way down from the 10^{19} GeV limit originally proposed, then that would be a tremendous, unanticipated bonus from the LHC. One can even say that this one result alone would pay for all the expenditure and the effort of building the LHC.

Let us take minute off to digest all this heavy stuff. We are saying that there are $n+3$ spatial dimensions to start with. The n extra dimensions then compactify into circles of radius R, where R is small like a millimetre perhaps, or may be a bit smaller even. This figure of 1 mm is extraordinarily large compared to what people used to talk about earlier in the context of compactification. When they talked of compactification, they always had in mind a R value of something like 10^{-33} cm or so, i.e., a length of the order of Planck length. Correspondingly the energy involved would be very high, of the order of Planck energy. But if R is like one mm or so, the string length could be as high as 10^{-17} cm, and this would bring down the energy to the TeV level, making it just about possible to explore string physics with LHC rather than throwing up our hands saying, "Oh we need an accelerator as big as the Cosmos! Forget it!!" If you recall, this is what I did in Volume 1, but now the picture seems to have changed!

Here is some more of this heady stuff! Look at Fig. 17.5. What we see here is a schematic of our universe (in 3-D). What is conveyed is this: The three forces, the weak, the strong and the electromagnetic act in 3D. But gravity is different; it can 'leak' out to the extra dimensions even though they are curled up; and this leakage can produce bizarre effects. In more technical terms, the physical 3D Universe that we live in is a D-brane. The additional dimensions curl around the D-brane. This is point number one. Next, electrons, photons, W, Z, etc., are tiny strings the two ends of which are stuck to the D-brane. Gravitons, on the other hand, are closed loops. They are therefore not stuck like electrons etc., and can merrily wander off into the extra dimensions.

You may think that all this is a lot of hocus pocus. Figure 17.6 should assure you that it is not. You may reluctantly agree with the picture, and yet wonder what is the big deal in it. Just this. Our Universe is merely one portion of something much bigger. That something that is bigger is believed to be like an infinite sheet that is folded like computer printer sheets. (These days one uses mostly ink jet printers and A 4 sheets, but barely a few years ago when we had to use line printers, this is how computer printing paper came. Even today, heavy-duty printers in many offices use this kind of stationery.) The overall picture is thus like what it is in Fig. 17.7. Our Universe is vast, and yet, there lies right next to it, 'barely one mm away or even less perhaps', a parallel Universe that we cannot see! We cannot see it because if light from it has reach us, it has to travel a long, long distance and therefore for a long, long time as Fig. 17.7 tries to illustrate. Nevertheless, a galaxy in this parallel Universe can interact with ours more directly through gravity because gravity, as already pointed out, can wander in all dimensions. By the way, this is one way of explaining dark matter. It is not there

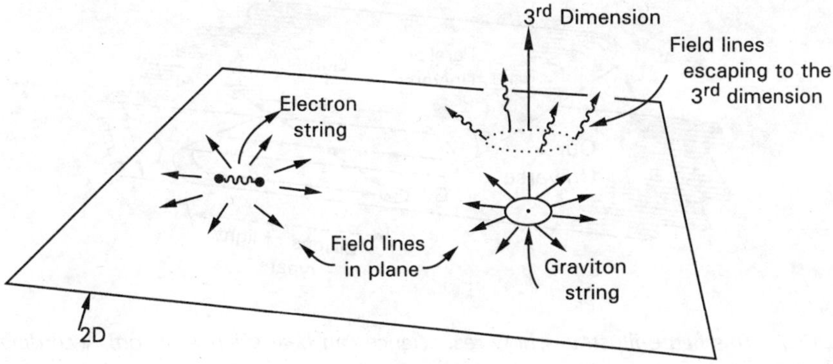

Fig. 17.5 *The sketches above illustrate further the confinement of forces other than gravity to three dimensions of space. For simplicity, we consider (2+1) D. The electron is a string with its two ends tied to the 2D space in which it exists (shown here schematically as a plane). The graviton, on the other hand, is a closed loop and can wander off into the third or the extra dimension. The consequence is that the electric field lines of the electron are confined to the plane while gravity lines can 'escape' into the extra dimension. When they do, their behaviour can be different; that is the essence of the previous figure.*

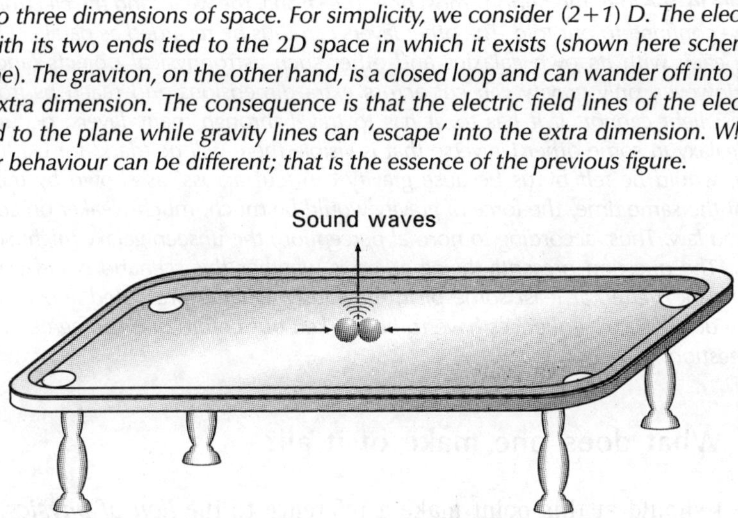

Fig. 17.6 *This figure offers a simple analogy to illustrate how while the graviton can escape into the extra dimension, the other strings (particles) are confined. Let us start with a billiard table. Say there are two billiard balls on the table and that they collide. The space of the balls is two-dimensional, and the trajectories of the balls are confined entirely to these two dimensions, i.e., to the surface of the table. However, when two balls collide, the sound that is produced (the acoustic waves) can escape into the surrounding air, that is, the third dimension. Future experiments at the CERN LHC could possibly reveal something like this. High-energy particle collisions could possibly produce mini black holes. These would emit Hawking radiation (see Chapter 13) which would be in our world. Gravitons also would be emitted but some of these might escape into the extra (curled) dimensions.*

in 'our Universe', but in another parallel sheet; that is why we can feel it but not see it! And because of the $1/r^{(2+n)}$ law that operates at short distances, this 'nearby' galaxy appears to be very far away. As far as I am concerned, this is just too much. What will they think of next?!

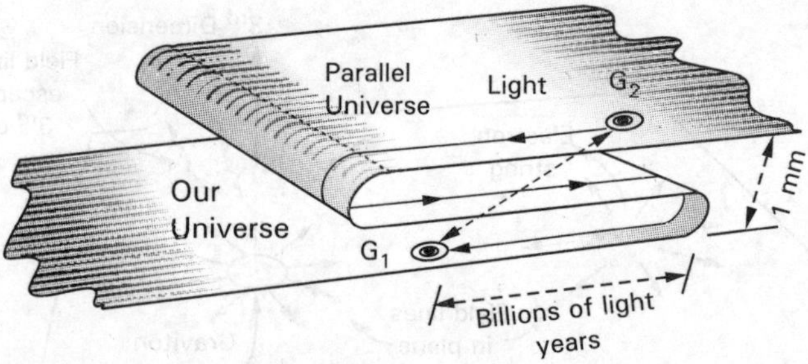

Fig. 17.7 *This figure illustrates how real science can beat science fiction! According the M-theory, our Universe with its three physical dimensions of space might be just one 'piece' of an infinite Universe that is folded over and over. Just two folds are shown above, and we are in one fold, the bottom-most one. The strong, the weak, and the electromagnetic forces are confined to our fold. The other layers can thus be regarded as parallel Universes, possibly each with its own galaxies and other such astrophysical objects (and people also?). However, only gravity can cut across extra dimensions and reach us from other Universes; light cannot. If it has to, it has to travel through many layers, or 'universes'. Thus, a galaxy in some other Universe that is visible there will be 'dark' for us. Its gravity, however, would be felt by us because gravity can 'cut across' as shown by the dashed arrow. At the same time, the force of gravity would be much, much weaker on account of the altered law. Thus, according to normal perception, the unseen galaxy might seem very far away. The question presently being asked is whether this scenario could explain the origin of dark matter, at least some of it. Obviously, what is presented above is only an idea. The detailed consequences have to be worked out before one can agree or rule out the suggestion.*

17.8 What does one make of it all?

May be I should at this point make a reference to the *lion of physics*, since I started with the lion. Physicist Michio Kaku says, "For the first time, we can see the outline of the lion. One day, we will hear it roar!" Hope so!

Another physicist says: "String Theory is smarter than we are. It knows what spacetime is, and we don't, and we have to figure out how to ask it!"

Yes, a lot of people are mighty excited about M-Theory. But there are other cautious voices also. For example, Steven Weinberg, one of the three architects of electro-weak unification, is more conservative. He says that no one knows as yet how to write down the basic equation of the M-Theory, like the Schrödinger equation or the Dirac equation in quantum mechanics. Remember the conversation between Dirac and Feynman on writing down such an equation for elementary particles? (See ***QR II***.) Weinberg adds that we have not even discovered the fundamental principle that would govern such an equation (such as wave–particle duality in the case of Schrödinger equation, and the action principle in the case of Feynman's version of QED). What makes the discovery of such a principle very difficult is that to arrive at a proper theory in four dimensions, we must start with 10, may be 11, or even higher dimensions. And those dimensions are so far removed from intuition we don't even know how we should go about guessing. Suppose we did discover the underlying principle and were also able to write down the equation. What is the guarantee we would know how to solve it? Right now most of our experience is with doing perturbation theory, which is OK if the coupling is weak. The forces of the fundamental theory might be neither weak nor strong, and we may be in the middle, in no-man's land. As a result we may not be able to seek the help of duality and get the answers. That is Weinberg! But Weinberg freely admits that suddenly, a young genius somewhere may make all this irrelevant by finding all the answers in a most unexpected manner, as early as tomorrow! This sort of thing has happened before and it could well happen again.

Weinberg is cautious and polite in his warnings about String Theory. But many others go much further in their criticism. They say (effectively), "What is the use of going gaga with just ideas? Till today, Super String Theory or its newer cousins have not made a *single worthwhile prediction* that can be put to a stringent experimental test." One of the pundits of the earlier era, 't Hooft (we met him in Volume 1; he is now a Nobelist) says, "Coming as they do, these reports from the string clan are as usual loaded with overstatements. ...String theorists tend to trust their theory blindly, claiming it can deal with everything. In reality, they don't understand gravitational collapse any better than anybody else."

So what's going to happen? Will string/brane theory deliver the goods or fade away with a whimper? You can lay your bets; I will just watch!

17.9 Summary

- First there was just the plain String Theory. With SUSY incorporated, it became Super String Theory or ver. 2.0 of String Theory.
- Then came duality, Asoke Sen's seminal paper, and Witten's inspiring leadership. A new wave started and, using duality, many apparently different versions of the String Theory could be inter-related.
- Meanwhile, people went from strings to membranes or branes for short, and this raised dimensionality from 10 for String Theory to 11 for M-Theory.
- Branes are shedding some light through mini black holes on the topological nature of spacetime immediately after the birth of the Universe, when it was believed to be something like a 'foam'.
- For making contact with the real world, one has to come down from 11 or more dimensions to 4. This is done via compactification.
- Compactification has raised the interesting possibility that there could be departures from the traditional inverse square law for gravity at small distances of one mm and less.
- Such departures also have consequences for the nature of the Universe itself, allowing gravity to escape into the dimensions beyond four while other interactions are restricted to the 4-D spacetime in which we find ourselves.
- While the fans of the string and brane theories are all excited about the possibilities that these theories seem to hold, there are others who are not so sure twhether these 'star' theories would finally deliver something concrete. Will they (the star theories) fade away without a whimper? Time alone can tell.

18

BRINGING THE CURTAIN DOWN

This chapter wraps up the subject of the micro- and the macrocosm, with which we have been concerned in these two volumes. The two chapters that follow wind up the entire Vignette Series.

Now that we are almost at the end of the grand cosmic tour, how do you feel? Rather dizzy? I wouldn't blame you, for I feel that way myself! When I came out of college, parity violation was still to be discovered, neutrino was just heard of and no one had ever caught it yet, the laser had not been even conceptualised, and spacecrafts still belonged to science fiction. Today, all that has changed.

All the quarks have been detected, the neutrino has been trapped, in fact neutrinos from supernova explosions have been detected, the W and the Z have been added to the family of elementary particles, there are telescopes in space, and particles have become old hat giving place first to strings, followed by D-branes, etc. I mean what's going on currently is simply mind-boggling. I can say with certainty that way back in 1950 or even in 1960, no one, and this includes all the greats like Dirac and Heisenberg for example (who were then still alive), could have foreseen such developments in their wildest dreams. Right now, so much is going on that I simply cannot tell you everything about it. Thus, what I have offered in this volume is but a small sample. And to emphasise that this is only the tip of the iceberg, I have included below a few sections that briefly deal with some of the stuff that I have had to leave out in the earlier chapters. The list of topics dealt with below is by no means exhaustive; it is a mere sample to indicate that I have really not been able to cover the whole spectrum—it is too vast anyway, and constantly changing.

18.1 Towards gravitational astronomy

In Chapter 6, I made a brief reference to gravitational waves and their detection. Serious efforts are now underway to use gravitational waves to open yet another astronomical window. I shall now describe some aspects of this work currently in progress.

Let us start by comparing the familiar electromagnetic waves with the less familiar gravity waves. Electromagnetic waves are caused by moving electric charges. In most of the cases we normally encounter, the motion is that of electrons in atoms and molecules. By contrast, gravity waves are *produced by the motion of spacetime itself.*

Next, in the Cosmos, the sources emitting electromagnetic radiation are huge compared to the wavelength of the radiation. As a result, one can image the source (this is what one essentially does using a telescope). In the case of gravity waves emitted by astrophysical objects, the wavelength is often comparable to the size of the object.

In their passage from the source to the earth, electromagnetic waves coming from outer space get absorbed easily by objects that intervene. On the other hand, since gravity waves interact feebly with matter, they suffer very little attenuation in their passage to Earth. However, this advantage is also a disadvantage since this weak interaction makes it so much harder to detect gravity waves!

In their lowest order, electromagnetic waves are dipolar. Gravity waves, on the other hand, are quadrupolar in their lowest order. They act tidally, stretching and contracting any object they pass through.

Turning now to the sources of gravitational radiation, where the Cosmos is concerned there are two basic sources of gravitational waves that one can think of. Quantum fluctuations associated especially with the inflationary era (see Chapter 12) form one source, that is to say, the entire Universe is itself the source. However, these waves associated with the primordial fluctuations of the Cosmos would have been travelling for nearly fifteen billion years or so and therefore also over a very long distance. Thus their amplitude would be very weak, and we just cannot detect such waves in the present time. However, we can detect their footprints in the cosmic microwave background, as was explained earlier. What about the second source? Basically, the second category is made up of exotic astrophysical objects and events such as colliding black holes and supernova explosions, for example.

At present, six gravity detectors are in the process of being commissioned in various parts of the world. All of them use some form of laser interferometric principle, since that gives a very high sensitivity of detection. The details are complicated and I am afraid I cannot go into them here. I shall, however, give a brief description of the three American detectors. Each of them basically has two (very long) arms at right angles to each other. Let us say that one of these arms is aligned with the x-axis and the other with the y-axis. Suppose now a gravity wave comes down the z direction and hits the detector. It would then slightly warp the two arms; in fact the warping would be oscillatory in nature, even as the gravity wave oscillates. As a result, the interference pattern generated in the interferometer is disturbed with respect to what it was when there was no wave. What the instrument must do is to detect this tiny oscillatory warp caused by the impinging gravity wave.

All this is more easily said than done because the warp is incredibly small. How small is this warp? You wouldn't believe—10^{-21} cm, yes that small! As someone put it, it is like "noticing that the Saturn has moved closer to the Sun by the width of a hydrogen atom." Concerning this incredible objective, one of the authors of the proposal for this fantastic detector said, "The prospect of achieving such sensitivity often strikes one as insane." But, believe it or not, today's technology is on the verge of achieving such sensitivity!

Technology is no doubt capable of great triumphs, but these do not come easily. High sensitivity means that one must pay very careful attention to all the possible sources of noise and minimise noise to the maximum extent possible. For example, atoms and molecules in the mirrors and support systems would be undergoing thermal vibrations and this can generate noise; that is one reason for trying to detect gravitational waves of low frequencies because at these frequencies thermal noise would get averaged over. (Basically, over a period of time that relates to the period of the gravity wave, the billions of positive and negative excursions due to thermal noise would cancel each other.) The laser power too affects observations. Too little power means the light signal would be weak and can introduce its own source of errors, while if the laser power is too high it can kick the mirror around. The tube in which the laser light travels is evacuated to a high degree. Even so, the tiny amount of residual gas can cause minute fluctuations in refractive index, producing its own source of noise. Perhaps the noise that is most difficult to eliminate is the natural as well as man-made seismic noise. I mean, there are constant movements of the Earth due to the movements of the plates below, due to waves lashing on the nearby beaches, movements of trees in the neighbourhood due to strong wind, tornadoes, etc., not to mention man-made noise due to trucks moving about and so on. Even if these events are taking place some distance away, their effect can be quite serious. In short, noise reduction and vibration isolation become *exceedingly* important and have almost been made into a fine art where these gravity detectors are concerned.

OK, where are the six gravity detectors located? Three are in America, two in the State of Washington and the third in the State of Louisiana. How much does the American project cost? A cool half a billion dollars! The American detectors are called LIGOs, standing for Laser Interferometer Gravitational-wave Observatories. Getting funding for the LIGO project has been far from easy. When the proposals were put up, some people said, "Good idea but it can wait because it will not work right away. We cannot detect the signals promised because the events are expected to be rare. We can use this money much better in other ways." Kip Thorne, one of the Godfathers of the project countered, "Theorists have a poor track record for predicting what we will see when a window is opened. Early radio telescopes discovered that the signals we . much stronger than what the theorists had predicted. That happened again when the X-ray window opened in 1960's. And when we started looking for neutrinos coming from the Sun, we were surprised by how few there were. In a similar manner, opening the gravitational window would give us a more radically different view of the Universe than those previous did." A strong defence! Finally, the project did get sanctioned.

Getting back to the detectors, I said there are six and that three of these are in America; which and where are the other three? TAMA 300 in Japan, GEO 600 in Germany, and Virgo in Italy. Compared to the American detectors, these are much more modest. Can they then compete? To a certain extent yes, because when money is short, ingenuity comes to the rescue! But all in all, it is a case not of competition but cooperation. As one LIGO scientist puts it, "If both the LIGO sites (in Washington State and Louisiana) see the same shape signal within a few milliseconds of each other and so does a GEO sitting on a different continental plate and connected to a different electrical grid, then it (the signal) is very, very unlikely to be a fake signal from some common source of noise." Further, since the gravitational wave would hit different detectors at slightly different times, one can (guess how?) pinpoint the direction from which the gravitational waves came. In other words, while the American detectors would be able to say a lot about the detailed signature, the others would help in confirming the genuineness of the signal besides identifying the direction of the source.

Interestingly, even before significant results are available, physicists are already planning improvements and the next generation detectors! The Americans expect to upgrade the LIGO to LIGO II. Basically, one wants to go as low a frequency as possible. And in 2010, NASA and the European Space Agency expect to launch three satellites that would fly in a triangular formation. This system called LISA or Laser Interferometric Space Antenna would be able to detect gravitational waves of very low frequency, of the type that one cannot ever detect on Earth due to seismic noise. So, look out for exciting announcements concerning gravity waves! That news would sure make waves!!

18.2 Quark–gluon plasma

Here on Earth, the highest density matter one encounters is that of the insides of the nucleus. (Just to get a feel, calculate the density of nuclear matter. Take, for example, the uranium nucleus, ^{235}U say. Find out its radius from somewhere or the other. You know the nucleus has 92 protons and 143 neutrons. With this information, calculation of the density is just high school arithmetic. But the answer would be revealing when you compare it with the density of ordinary matter.) The question arises: "Can there be other phases of matter with still higher density? Where do they exist, if at all? Were any of these phases present during the early history of the Universe?" These questions are not quite academic, and have in fact been triggered in part by studies on QCD.

Let us look around a bit. Consider a neutron star. As I have told you elsewhere, it has a diameter of about 10–20 km, and a mass several times that of the Sun. If you calculate the density of a neutron star, it will be astronomically high. What this means is that right now, there exists out there in space, matter with much higher densities than we know of on Earth.

This raises the question: "Can we have a phase diagram for 'cosmic' matter, say a diagram that shows the different phases of matter pertaining to high-energy physics as a function of temperature?" People have for some years now been asking this very question, and not only are there some interesting ideas, but also many experiments have been proposed to verify some of these concepts.

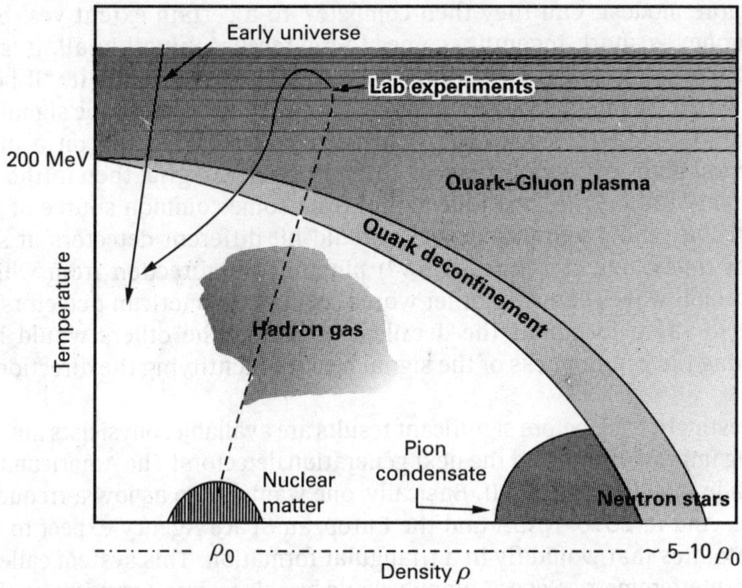

Fig.18.1 *This is a highly schematic phase diagram of matter that covers astronomical range of densities and temperature. In this case, the density of nuclear matter (denoted by ρ_0) is a convenient unit. If you will recall from* A Hot Story, *truly speaking we must also take note of pressure, but we drop that from consideration here and consider just the T–ρ plane. The slanting arrow at the top left is a guesstimate of the trajectory that Universe might have taken during its cooling phase, immediately after the Big Bang. What we see from this is that it most likely went through a quark–gluon plasma phase and that is why there is so much interest in this particular phase. The curve marked Lab experiments shows how by exciting matter via relativistic nuclear collisions, one expects to momentarily raise matter to the quark–gluon phase, giving us a glimpse of the same.*

Figure 18.1 shows schematically a diagram of the type just mentioned. As you can see, at one corner almost we have a state of matter that we can call nucleonic, i.e., the stuff of which ordinary nuclei are made. This is liquid-like, and it is thus that a nucleus is often compared to a liquid drop. If the temperature is raised, then this 'liquid' evaporates and one gets what is termed a *hadron gas*. On the other hand, if, keeping the temperature low one increases the density, one obtains a solid-like phase where one has a lattice of pions; this is called the *pion condensate*. Just around this region we also have the type of matter of which neutron stars are made; the temperature is low but the matter is very dense as I have already explained in ***Chandra***.

You will notice in this figure a curved band, almost like a quadrant of a circle; this band is important and requires comment. To appreciate this comment, consider Fig. 18. 2. This shows nuclear matter in which there are protons and neutrons, each with three quarks. It is as if the quarks in each nucleon are in a 'bag'. Suppose this nuclear matter is squeezed, that is to say the pressure is increased; for good luck let us suppose the temperature also is increased. What happens

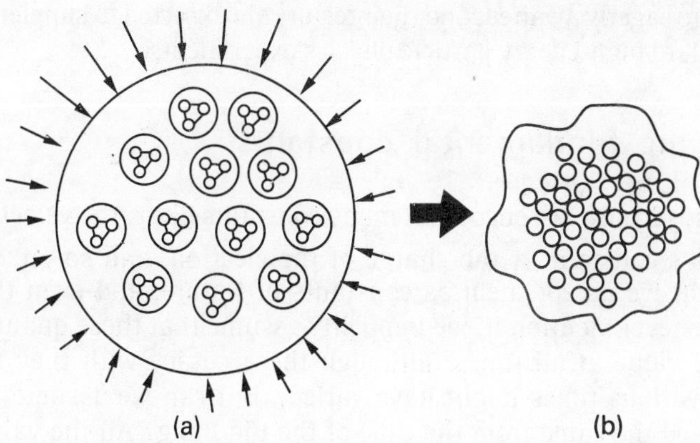

Fig. 18.2 *This figure seeks to shed more light on what the high-energy collisions seek to do. Consider a nucleus. It has protons and neutrons each of which is made up of three quarks confined in their own respective 'bags'. When one smashes two very high energy beams of say lead nuclei, the collision generates momentarily very high temperatures and pressures which break the 'bags' and fuse the quarks all together into quark–gluon plasma (QGP). Of course, this state would disappear quickly, but while doing so would also leave behind remnants. It is by looking at those signals that physicists expect to get some idea of the QGP phase of matter.*

then? If the temperature and the pressure are *substantially* increased, then a stage would come when the quarks brake out of their respective 'bags' and merge, forming what is called *quark–gluon plasma.* Sometimes, people refer to this as *quark matter.*

OK, but how do we know this is true? From a theoretical point of view, there are models that strongly suggest this. The curved band in Fig. 18.1 divides quark matter from the other phases—it is not quite sharp like the phase boundary we are used to in condensed matter physics (see ***Many Phases*** and ***A Hot Story***). Perhaps that is because of the theoretical uncertainty that surrounds such speculations at the present time. Basically, it denotes a region over which *quark confinement/deconfinement* occurs.

Turn now to some of the arrows and lines in this figure. One of these lines marks a trajectory that presumably represents the way the Universe evolved while cooling after the Big Bang. This suggests that at some stage the Baby Universe must have been composed of quark matter—that is the reason for the interest in quark matter. But how can we be sure that such a phase of matter is possible? This is where the other arrow comes into the picture. It shows the conditions that are expected to be reached in certain high-energy ion–ion collision experiments currently under planning in Brookhaven in America and at CERN.

Basically, in these experiments heavy ions (like lead, for example) would be accelerated to energies of the order of 500/*A* GeV (here *A* represents the atomic number).Two such beams coming from opposite directions would be allowed to collide head-on, and this, it is expected, would tear the quark bags fusing the quarks into a quark phase, at least for a short while (as shown in Fig. 18.2). These experiments are eagerly awaited, and their results are expected to supplement those from the standard high-energy particle physics experiments.

18.3 Varying fundamental constants

When we study physics, we encounter many constants like c, the velocity of light, $\hbar$, the Planck's constant, e, the charge of the electron, and so on. Without a second thought we accept them as constants of Nature, and even though the teacher might never mention it, we implicitly assume that these quantities have had the same value at all times (although the accuracy with they have been measured at various times might have varied). But can we assume that these values remained the same from the time of the Big Bang? All the various ideas presented in the earlier chapters (including, for example, Linde's idea of chaotic inflation), have tacitly assumed that the constants of Nature have retained their values without any change right from the time of the Big Bang so to speak. But how do we now this is true? We do not!

Many decades ago Dirac wrote a paper in which he argued that G the gravitational constant varied with time, being inversely proportional to the age of the Universe. Despite his eminence, few took this idea of Dirac seriously. Meanwhile, here and there some experimentalists tried to check whether there was any experimental evidence for changing constants. There were none, going as far back as 4.6 billion years. But what about still earlier times? No one knew for sure.

Recently, John Webb and his colleagues have been studying the radiation coming from distant quasars. I suppose you know that quasars stand for quasi-stellar objects. They are compact but massive, very far away, and radiate an *incredible* amount of energy. People think that perhaps quasars contain black holes at their centres, and that the immense gravitational force exerted by these black holes converts matter in the vicinity very efficiently into light. As this light from distant quasars travels to the Earth, it gets absorbed to some extent by the galaxies en route. Thus, the quasar light is rather like sunlight, with dark bands, Fraunhofer bands if you wish. Or, as some one put it, it is the 'bar code' of the galaxy! This 'bar code' would give the value of α, the fine-structure constant of those times. This way, one can go back to times when the Earth and even our Sun did not exist. Thus, by comparing the 'bar codes' for quasar absorption spectra with similar 'bar codes' of atoms and ions in the laboratory, we can say whether the value of α has remained the same or not over billions of years.

OK, so what do these quasar studies say? That you can see in Fig. 18.3. What is shown here is the fractional change ($\Delta\alpha/\alpha$) as a function of z the redshift, which, remember, is a measure of the distance of the object from us, and therefore means going back in time so to speak. The time span ranges from about 5.6 billion years (for $z = 0.55$) to 12.6 billion years (for $z = 3$). Each data point is the average of over ten measurements. If α was not changing, the data points would be scattered randomly above and below the x-axis. Instead, they appear to lie systematically below for z greater than a certain value.

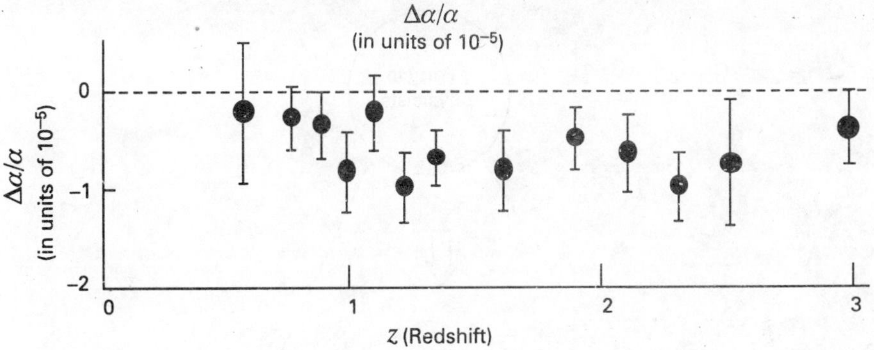

Fig. 18.3 *This figure presents a summary of the data on quasar redshifts, as obtained by Webb and his associates. Each data point represents an average over ten independent quasar measurements. We see that while for small z the fractional change in α is zero within experimental errors, the value of $(\Delta\alpha/\alpha)$ for higher values of the redshift is consistently below zero, beyond experimental errors. This is the basis for wondering whether the constant has been changing with time.*

Are we to believe this, and if so, what do we make of it? As far as the first question is concerned, in physics one would always like confirmation of experimental results, especially when they imply something revolutionary like this data seems to. Many think Webb's finding is correct but would like it to be confirmed by independent experiments. The chances are that the results would be confirmed. In which case, what do they mean, and what are their implications? I am not too sure, but one thing *is* certain: confirmation that α is changing would have a profound effect on physics and in particular on our understanding of the early history of the Universe. So, keep an eye on the α-story!

Well, that brings me to an end of the additional topics I wanted to give you a glimpse of. There are of course many more things I could have added, but as I said earlier, I am here just giving a sample of the many exciting things that are happening that I had to reluctantly pass. You know something? A good fraction of the advances I have described in this volume are due to people who were born in the second half of the twentieth century, and were fairly young when they helped physics to take big leaps forward. Before I say more on that, let me call attention to one important point.

18.4 A grand convergence?

In this volume we have essentially been focussing on the birth of the Universe (or should I say our particular Universe?) and its very early history. Basically, three groups have been involved in this reconstruction of the early history of the Universe. They are: (1) astrophysicists who are all the time looking out into space and trying to make sense of what they see, (2) particle physicists who want to describe physics in terms of strings, membranes, hidden dimensions and what have you, and (3) lastly the cosmologists who are always interested

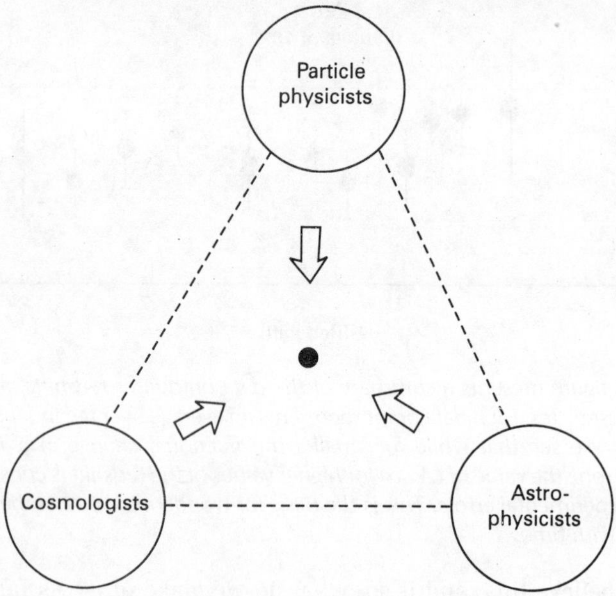

Fig. 18.4 *This figure schematically shows how three categories of physicists, the astrophysicists, the particle physicists and the cosmologists are all marching towards the same goal. The important point is that they are not marching independently but by holding hands. It is that intensive cooperation that would one day reward them with success. This interdisciplinary cooperation is one of the most satisfying aspects of modern physics.*

in the physics of the Universe as a whole instead of just one bit of it as other physicists often are.

Let us imagine these three groups to be at three corners of a triangle (see Fig. 18.4). Their goal is the centroid, and they are closing in from three sides, keeping in constant touch with each other. This is the best way of describing what's going on at present.

Just to remind you about what we know at present:

- We feel pretty sure that the Universe originated in a Big Bang.
- We can say with certainty what happened one second after birth.
- Before one second, we cannot be that sure. May be we can go up to 10^{-12} s, but beyond that the picture continues to be hazy in many places. However, we do not expect it to be that hazy for too long.
- We definitely know that the Universe is expanding right now. About the future, it all depends on Ω. After a long period of uncertainty, the omega situation appears to have become somewhat clear. I think we can safely bet that Ω is close to 1.
- We may not know many of the details, but it sure appears that the Universe went through a tremendous expansion around $t = 10^{-35}$ s. This is called inflation, and is considered by many to be the Big Bang itself.

- Many believe that although inflation is acceptable, the version of inflation that is probably correct is the chaotic inflation of Linde.
- Again, though the details are fuzzy at present, it is more than likely that we have to describe the early Universe via some version of the M-Theory, complete with multi-dimensions, compactification and all that.
- Right now, ideas such as parallel universes, etc. are mostly speculation. Possibly, some of these may come out to be true while others may fall by the wayside.
- We are still waiting to catch the proton in the act of decaying. Will we ever detect a clean decay event? I don't know. Maybe all we would be able to do is to push further the supposed half-life.
- We are yet to have a GUT that is accepted by all without question, like the Salam–Weinberg–Glashow theory of electro-weak unification.
- After that, this GUT has to become a seamless part of whatever theory that emerges, that which makes a smooth unification with gravity, possibly a Brane-Theory of some kind in a suitable number of dimensions.
- Almost certainly, the brane-picture, when final, would tie up the strong force and gravity in a quantum framework.

I think I have said enough to indicate that we are now in a most exciting period in the history of physics, and for the first time we can even seriously wonder whether physics might possibly come to an end.

At the same time I want to caution that the curtain has not come down yet. There has been no shortage of false alarms. Many, including some of the greats, have declared in the past that it was all over, bar the shouting, but Nature has repeatedly fooled them all! (Remember what Dirac is supposed to have said? "The rest is all chemistry" or something like that. What tremendous progress has taken place since then!)

Let me at this point quote what Weinberg has to say. He observes: "The discovery of a unified theory that describes Nature at all energies would put us in a position to answer the deepest questions of Cosmology: Did the expanding cloud of galaxies that we call the Big Bang have a beginning at a definite time in the past? Is our Big Bang just one episode in a much larger Universe in which big and little bangs have been going on eternally? If so, do what we call the constants of Nature or even the laws of Nature vary from one bang to another?"

Weinberg adds that even if we are able to answer these questions, it would not mean the end of physics. 'Conventional' physics by which I mean physics related to solids, turbulence, etc., would go on for a long time. But the discovery of unified physics would certainly mark the end of one kind of physics, the end of a kind beyond which there are no further layers. To this astrophysicist Martin Rees adds that today, physicists are tackling three basic frontiers—the very small, the very big and the very complex. Rees adds, "A star is simpler than an insect. The fierce heat within stars and in the early Universe breaks everything down into simple constituents. It is the biologist whose role it is to explain intricate, multilayered structure of trees, butterflies and brains that faces the tougher challenge." What this means is that even after Cosmology and particle physics come to an end (if they ever do), that is to say the story of the big and the small is over, there will still remain the very strong challenge of dealing with highly complex systems. In fact, there is already a subject known as *complexity*, but for obvious reasons I have maintained a healthy distance away from it!

What I want you to presently appreciate is that it is risky to declare that the game is over! Quite possibly, there would be wonderful physics to do for a long, long time. And that should give a lot of you youngsters much hope. In other words, the progress to the 'centriod' could well be asymptotically long! Know what I mean? (A favourite expression of Raman!)

18.5 Closing remarks

Challenges are often compared to the Everest. Frontier physics is undoubtedly a great challenge but it is very different from the challenge that the Everest poses. In the case of Everest, when Tenzing and Hilary first scaled it, a lot was already known about the challenger. It was therefore largely a case of careful planning, good training, proper conditioning, and guts as well as grit. But in our case the story is vastly different; we know very little about what to expect except that there can be surprises galore!

There is also another important difference. When one reaches the summit of a mountain, one knows for sure that one has arrived. There is no more climbing to do, and parts of the landscape hidden till then by the mountain now come into view. For example, when Tenzing and Hilary went up, they were told, "Once you reach the top you would be able to look down into Tibet and China; that, you cannot do from here. When you can see down into that plateau, then it means you have reached the top". In the case of physics, I am not at all certain that one can ever assert that the end of the road has been reached. In the case of a mountain, we know for sure that there is nothing beyond the summit, but in this case, i.e. of the very big and the very small, how can we be so sure? I don't know if others would agree but that is the way I feel. But this much I do know:

- The journey, if it continues endlessly, would always be exciting; it was so a hundred years ago, and it is so now also. So there is plenty in store if you want to join.
- True, some aspects of the studies ahead are difficult. But history has invariably shown that somehow, the young are able to cut corners and put the finger on the button. That is what happened with Einstein; that is what happened with Heisenberg; and it has happened that way again and again, many, many times. Being young is therefore not a handicap but a blessing. I thought I should specially emphasise this.

- About toughness next. All are not made the same way. Some can be very comfortable with duality, D-branes and all that. But others may be cut out for building neutrino telescopes under the sea, developing computer programs to analyse complex data, and so on. In short, there is work for people with all kinds of background, skills, and aptitudes. Don't hurriedly disqualify yourself and drop out! Be brave and venture forth!

Good luck and happy hunting when you get into the game.

19

EFFECTIVE THEORIES AND MOTHER OF THEM ALL

Well, dear reader, it is almost time to say the final goodbye but before I do so, I would like to draw your attention to a few important points in this and the following chapter.

In this series, I have tried to take you with me to a sea shore, showing you a few pebbles here and shells there, trying to arouse your curiosity, and encouraging you to dive into the adjoining ocean to look for pearls. If your vision is broad enough, and if you are lucky, pearls you will surely find!

What do I mean by broad vision? To explain this, let me remind you how science is taught in our country in schools and colleges. We have subjects like physics, chemistry, biology, and so on, and these are often taught as if each of these subjects had little or no connection with each other. No, these subjects of science are *not* non-interacting and independent universes! Truly speaking, science has no borders; only humans create and impose them. Once upon a time, people said physics is physics and chemistry is chemistry. Yet we have now physical chemistry, and chemical physics! (Try to find out what is the difference between the two!) In fact, there are any number of such hybrid or cross disciplines. What this shows is that while boundaries may be drawn for operational convenience, truly speaking, Nature knows no boundaries. I mean, just look at the human body. All sorts of sciences are in action there, from electro-chemistry to neural networks, not to mention such things as mechanical engineering and control engineering! Yet we have universities in India that tell you , "Your basic degree is in chemistry; therefore your Ph.D must also be in chemistry; we cannot allow you to switch to physics", and so on. To put it simply, God knows no boundaries and does not have hang-ups like we often do.

Having said all this, I must mention that at the practical level, even within physics, it is prudent to recognise the existence of 'layers'. Let me explain. A few hundred years ago, people believed that matter was homogeneous and continuous. Today we know this to be untrue; yet, the calculations they made in those days with this seemingly wrong assumptions about elastic structures, etc., turn out to be correct. In fact, even today, structural engineering is taught as if matter is homogeneous.

Go down one level and enter the world of metallurgy. Here, one cannot afford to ignore the atomic nature of matter and such things as dislocations, grain-boundaries, etc, become very important. In case you are not quite familiar with these terms, let me mention that these are various types of defects in crystal structure. Even the presence of impurity atoms is important; not merely their presence, but also their location. The point to be appreciated is that the strength of materials and their failure are both dependent upon the various types of crystalline defects that happen to be present. To put it differently, at this level one can no longer ignore the atomic nature of matter.

In metallurgy we do not generally bother about the fact that the atom has a structure of its own, that is to say, it has a positively charged nucleus with electrons orbiting round it, but the presence of atoms we must recognise. When we descend from metallurgy to the physics of materials, the structure of the atom can no longer be ignored. It is only when the electronic structure of solids is properly dealt with using quantum mechanics that one is able to design transistors, solid state lasers, etc.

Go down one step and enter the world of nuclear physics. Here, one can no longer regard the nucleus as a point with mass A and electric charge Ze, where Z is the atomic number. One has to consider carefully the fact that the nucleus is made up neutrons and protons, held tightly together. At the same time, we do not bother about the fact that the neutron and proton are made up of quarks. In fact, tremendous progress in nuclear physics had already been made before Gell Mann came up with the notion of quarks.

I suppose you are getting the general idea of what I am driving at. There are layers upon layers, each layer representing a theory dealing with matter on some particular scale of length and energy. Further, each layer represents an *effective theory* that provides support for the layers above and rests on all the layers below it. In the lowest layer (which is what we have been mainly concerned with in these two volumes) we are interested in very small distances and the correspondingly very high energies.

Let us call this lowest layer as level 1, and climb up from there to the next higher layer, that is layer 2. We can function quite comfortably in layer 2 without knowing much or even anything about the layer below. How come? This is because the basic physics of layer 1 enters that of layer 2 usually in terms of some unknown constants whose values are usually obtained from experiment. I mean, when we study astronomy we do problems taking the value of G from experiment; we don't bother about how G got its value, do we?

Thus, while at layer 2, even though we know nothing about layer 1, we can still get away most of the time because nowhere would we be dealing with distances or energies typical of the layer below. This happens layer after layer as we go up. Historically, however, we have not been going up; instead we have been coming down (in the process, enriching our knowledge). And as we came down, generally speaking, we never had to completely throw away our older pieces of knowledge. I mean things like Boyle's Law and Hooke's Law are still valid and still taught!

This layered structure of physics is something that is seldom stressed upon, though of course it is obvious to the pundits. No one told me about this, and I sort of had to discover it for myself as I went along. I want to save you that trouble and that is why I am drawing your attention to it explicitly. Having said all this, I must also hasten to point out that in certain scientific endeavours and projects, it becomes necessary to deal with many levels at the same time. Let me give you one example from my own background.

Take the case of a nuclear reactor. One builds a reactor to release nuclear energy in a controlled manner, after which one tries to make appropriate use of that energy. The 'heart' of the reactor is the nuclear fuel, and central to the design of the reactor are what are called reactor-physics calculations. Here one must deal with all aspects pertinent to nuclear fission and the chain reaction of the nuclear fission. However, in the process, one need not bother about engineering aspects such as say, whether the fuel rod is mechanically strong, etc. Now while the reactor physicist can afford to ignore many of the engineering considerations, the mechanical engineer cannot afford to do so. If the reactor physicist says, "The fuel rod must have a diameter of three centimetres, will be three meters long, will have natural uranium, will have a surface temperature of eighty degrees C, will be cooled by heavy water flowing round it in a tube of such and such a diameter, etc", it is up to the mechanical engineer to make sure that the engineering design meets all the requirements that the reactor physicist takes for granted. (Of course, the physicist is expected to use his common sense and not ask for impossible specifications!) Likewise, the nuclear metallurgist must make sure that the fuel rod is prepared in the appropriate manner, that the uranium has the correct density, is not spongy, and so on.

What I am driving at is that in modern science and engineering, we often have specialists in many different disciplines working together; each will be considering science at his or her own level as appropriate. This is happening increasingly, indeed even in many of the experiments that physicist are doing today to unravel the mysteries of the Cosmos. When Carlo Rubbia discovered the W and the Z particles, he was able to do so because van der Meer was able to give him the right accelerator for the purpose. And van der Meer was able to do his job because hundreds of people toiled in the background doing all sorts of things that were seemingly far removed from the electro-weak force. All this leg-work was vital, and without it Rubbia could not have gone to Stockholm.

Some day in the near future, you might perhaps read in the newspapers that the Higgs particle has been discovered or that a supersymmetric partner has been trapped and that this was done by so and so. Of course such discoveries are great for physics and the scientists who made these discoveries would no doubt deserve a big applause. But do not forget that these people would be standing on the shoulders of hundreds, literally hundreds, who toiled in their own respective spheres to make this kind of great achievement possible. And also remember that Nature has most kindly packaged physics and engineering in convenient layers, enabling people to function within their own levels and yet make important contribution to activities that cut across many, many layers. Spare a moment for these unsung heroes and heroines and applaud their silent work, even while admiring the new discovery and the spirit that drives such discoveries!

In summary, Physics is structured in layers, with each layer being appropriate for a particular scale of length and energy. All the layers except the bottom-most one represent effective theories. The one at the bottom of them all is what we are concerned with in this volume. It is not an effective theory, but the ultimate theory—that ultimate, fundamental theory—alone can provide the correct clue to the origin of the Universe.

20

SOME PERSONAL AND PHILOSOPHICAL REFLECTIONS

This final chapter is bound to evoke diverse reactions from readers. Young readers may be totally surprised by its contents. Among the older set, some may agree with me while many are likely to be displeased and even regard my words as a disservice. Each reader is welcome to his or her opinions; as far as I am concerned I feel I owe this chapter to myself. I do not defend its inclusion nor do I feel apologetic about it.

Almost towards the end of his famous book *A Brief History of Time*, Stephen Hawking says,

> We find ourselves in a bewildering world. We want to make sense of what we see around us and to ask: What is the nature of the Universe? What is our place in it and where did it and we come from? Why is it the way it is?

I find these words to be very significant. A long time ago, ancient man asked himself very similar questions. He too wondered:

> What is all this I see around me? Where did they all come from? Who am I? Where did I come from? How am I related to all this? And what is the role I am expected to play while on earth?

People in different parts of the world have always wondered in this fashion, and what we call knowledge is nothing but the synthesis of the answers people have been finding over the ages, in bits and pieces of course.

Tremendous progress has been made during the twentieth century in the advancement of scientific knowledge. None can dispute that. And Hawking, as everyone knows, has come up with some astounding and breath-taking ideas. He is, today, easily the best-known scientists in the world. I believe that Hawking's personal popularity is rivalled only by that of Einstein. And if Hawking asks the same questions as ancient man did (or at least some of them), we must pause and reflect, and that precisely is what I shall be doing in this chapter.

It is said that man cannot live by bread alone. In the same way, I submit that while passion for scientific enquiry and all that is most commendable, this passion must be suitably *tempered by various factors that are related to the society we live in*. By way of backing up this statement, let me quote what Victor Weisskopf (we met him first in ***WATT***) once said while addressing the members of the American Association for the Advancement of Science. He observed, "*Knowledge without compassion is inhuman, while compassion* *without knowledge is ineffective*." For us in India, this statement is doubly important. Why? Let me explain.

Close to sixty years ago, I was drawn irresistibly to science, for which two things were responsible. The first of course was the intrinsic charm of science itself, and the second was the call of Jawaharlal Nehru, the first Prime Minister of independent India. Nehru had always felt deeply about the problems of India, and he passionately believed that science and technology alone could solve India's problems. He missed no opportunity to air his firm faith in this magic recipe, and I was bowled over by his formula, as much as I was by science itself.

Nehru did not talk without a basis. The mighty Soviet Union (now existing only in history books!) had transformed itself from a miserably poor country into a very powerful one through massive use of technology. Technology is born of science, and that is why Nehru placed enormous faith in science and technology. Nehru was not alone in admiring the Soviet model. Saha [see ***Saha***] did so, Mahalanobis did so [see ***Saha***, chapter 1], and Bhabha [see ***Bhabha***] did so too in his own way. No wonder Nehru gave tremendous importance to science and technology and personal support to scientists like Bhabha, Krishnan [see ***Raman***], Bhatnagar, Mahalanobis and others. Many countries became independent around the same time as India, but they have not progressed as much in science and technology (at least in some areas). While our progress is no doubt due to pioneers like Bhabha, we cannot discount the tremendous initial impetus given by Nehru.

More than half a century has passed since those days. What does the scoreboard show now? Well different people would answer this question in different ways but I would prefer to reflect on it by recalling a letter I received about ten years ago. This was from a young schoolteacher living in a village in Bankura district in West Bengal. I did not know this teacher and had never met him. Yet somehow he wrote to me. In his letter he said (as well as I can remember), "Sir, they say India has made great progress in science and technology but we see very little evidence of it here in our village. For the people in the cities, we simply do not seem to exist. I am very keen about science. Please tell me what I should do to promote science and how I can make the students of our village take interest in science." So on the letter went. The message was very clear.

The existence of an uncomfortably large gap between the urban and rural areas in our country is well known. This gap encompasses income, standards of living, educational facilities, water supply, sanitation, healthcare and so on. It is said that today there are about four hundred million people below the poverty line in India. I remember a book entitled *Our India*, published in the late-thirties/early-forties of the twentieth century. That book told us that the population of India (and India in those days meant all of what is now India, Pakistan and Bangladesh) was forty crores or four hundred million. Just think about it! The number of people in our country today below the poverty line equals the *total* population of *undivided* India in 1940! Not at all a very pleasant thought.

Some may argue, "Listen. This has nothing to do with science and technology. It is all the result of bad politics and rotten administration." True, politics and administration have to carry their share of blame. However, I feel that the root cause of the problem lies much deeper. *Basically, we as individuals have all become insensitive.* Let me explain myself.

At the outset, let me make it very clear that I do not expect scientists to drop whatever it is that they are doing, rush to the villages and try to improve them. Rural development is no doubt very important but it must be undertaken by many agencies working together in a coordinated manner. Physicists may not be directly involved in rural development as such, but nevertheless, *they DO have an important role to play in overall national development.* If this is done properly, the fruits of science can and would reach the underprivileged.

The first thing that our physicists (and indeed all scientists) must do is to set apart some time for schools, students and teachers. We have several hundreds of excellent scientists, and they have all come through *our* schools and colleges. If these schools and colleges did not exist, where would all these scientists of today be? But somehow, most Indian scientists tend to suffer from amnesia. We must rememberthat a tree cannot bear fruit unless it is nourished and looked after.

I have heard many scientists complain, "Oh, young people do not appear to be interested in science these days." That may be so, but I wonder what these people are doing about enthusing the young and making them aware of the wonders of science. In the good old days, people like Raman took great interest in meeting school children and talking science to them. I distinctly remember late Professor Dhawan (who as Director of the Indian Institute of Science, Bangalore and the Chairman of our Space Commission did so much to put our space programme on track) once telling me, "You know, Homi Bhabha once came to our college and addressed us. That was what motivated me to take to science."

Today, very few of the bigwigs of Indian science have any time whatsoever for students in our schools. Most of them find plenty of time to travel abroad, but ask them to visit a small school and give a talk, they would grumble about having 'so little time'. It is not just the individual scientist; our various professional bodies are no better. In the UK, the British Association for the Advancement of Science does so much for popularising science and telling the masses about the fruits of science. Here, I have not seen any body do any such thing. Abroad, so many nice books on science for the curious are published, but in India there are not many writers and very few sponsors either. It is a crying shame indeed!. People are ready to throw money on film magazines that repeat the same gossip week after week, but these very same people would think twice before parting with any money to buy a popular-science book. For a brief period we had a TV show on science but that has vanished. No science academy (and we do have a few of them) has bothered to do anything to improve the situation either.

People think that villagers do not need science. May be they do not understand pulsars. But there are things they do understand. I have seen ignorant farmers pick up a lot of information handed out by agricultural scientists. I have seen villagers pick up watch repair, auto repair, etc., without any formal training. In places like Gujarat, Haryana, and Coimbatore, there are very enterprising people who make all kinds of gadgets. Just imagine how much better their products would be if they were given suitable exposure to some basics of science.

There is one thing scientists of all disciplines can do but fail to do, and that is to give active support to our teachers in schools and colleges. There is in India a body known as the Indian Association of Physics Teachers (IAPT). I am a life member of this organisation and have a tremendous soft corner for it. I have watched how this body has striven with enormous dedication over twenty years or so to improve physics education in the country. We have so many eminent physicists in the country. I wonder how many of them even *know* about the existence of IAPT. How nice it would be if teachers were to get the benefit of the experience and knowledge that the researchers possess? In America they have the American Physical Society (APS) that is a very active body of research physicists. Everyone who is anything in physics is a member of the APS. This body even has student members. The APS, however, does not function as a snobbish body. On the other hand, it has *close* links with the American Association of Physics Teachers (AAPT). In fact, every year, there is a joint session of the APS and the AAPT. In India, one never sees such cooperation. This is not to say that there are no concerned people at all; there are some dedicated souls who toil hard but how much better it would be if all our top scientific bodies were to regard such interaction as a sacred duty that they owe to our society. I see many of our bigwigs lament regularly about the decreasing enthusiasm for science, but apart from writing a few articles and perhaps making a few speeches (to elite audiences, of course), precious little is done by most.

Basically, it all comes back to extreme self-centredness. Once a scientist achieves some success, reputation and status, he or she somehow becomes caught in the quagmire of, as they say, keeping up with the Joneses. International science is dominated by America and there are good reasons for that; we need not go into these at present. People want recognition and appreciation from the West. True, in some respects peer recognition is a legitimate aspiration. Indeed this is what drives most people engaged in creative work, be it writing, music or art. So it is with science also. But there is such a thing as a limit.

At this point, I would like to briefly recall a saga of sacrifice that I have personally witnessed. This is the story of a man, a few years older than me. When I joined the TIFR about half a century ago and was working in the wartime barracks there (this was before the present building was built), one day a man joined one of the groups and started working in an adjacent lab. He had just returned from America. He seemed a sound experimentalist. Three things struck me about him. The first was his deep love for his mother, a poor village woman who cared very much for her son. For his part, the son was ever so grateful to his mother for all the sacrifices she obviously had made, and there were many. Secondly, this man was not at all ashamed of his humble beginnings. Even after spending many years in America working for his Ph.D, he remained a villager at heart. Not for him any airs or flamboyance. Thirdly, he cared very much for science on the one hand and our people on the other.

After a couple of years this man left to join a university. I don't quite know the reason, but I suspect that he was not quite comfortable with the rather insulated atmosphere that our labs tend to have. Subsequently, he spent all his life doing his best to improve the design of lab experiments for physics students. He did not have a big budget but used ingenuity wherever he could. The pundits of the land might have ignored him totally, but this brave man, working almost all alone and with very little support, did his very best to help students in ordinary colleges in small towns. I applaud and salute him for doing this. The stability and the progress that our country has achieved are built on the sacrifices of small farmers, poor workers, and others like them. In his own way, this man made many sacrifices so that our students may learn physics, understand its basic principles, and also learn how to do an experiment properly. Very few of the upper echelons have bothered about such matters.

In life, we receive so much from so many. Among other things, we also receive much from society, though this fact is conveniently overlooked by most. It is therefore just and proper that those who receive must also give back at least a little. There must be something like paying income tax. If our society supports us through funding, travel grants and so on, then we owe something back. Here, people do clever self-accounting. They say, "Look, I have trained so many Ph.Ds. Is that not a contribution?" In my view, this is not enough.

Just look a simple fact. More that fifty years ago, I recall there used to be small companies in Punjab that made various kinds of scientific equipment. I believe many of them still exist. Has all our 'great' progress made any impact on them? Has any scientific body—and we have so many of them—made any serious effort to talk to these companies and help them to build better and more modern scientific equipment? Hardly. I can write chapter and verse on this topic but I shall not. However, I shall mention but one point to illustrate my concern. There is a lot of talk about all kinds of advanced work on lasers being done in our country. In fact, I have seen in one lab a laser that could drill through one centimetre of steel. Very impressive no doubt; yet with all such 'progress' including the use of lasers in surgery, in guiding bombs or whatever, we still cannot buy an Indian-made 50 mW laser for lab experiments on diffraction. This kind of laser is required literally in *thousands* for our schools and colleges, but somehow, forty years after the first laser was built in India, we still have to import this puny laser! This one fact alone speaks volumes.

I don't want to make a big song and dance of the basic point I am trying to drive at. But I would like to recall some aspects from Raman's life by way of underscoring my view. Raman *always* had time for students, and especially in his later life he regularly invited school students to visit him in his institute. He would not only give them a guided tour, an inspiring one I might add, but also talk to them about many things related to science. How many are prepared to do that sort of thing today? Raman did not have to do it but he felt deeply for the country and its future, which he saw entirely in the young, and correctly so.

I shall end my musings by telling you about an incident that conveys in a nutshell the character of this great man. You know about the Bharat Ratna, the top honour bestowed by our nation, don't you? Well, Raman was in the first batch of Bharat Ratna Award winners. The award ceremony was to take place in the last days of January, soon after the celebration of the Republic Day. The then president, Dr Rajendra Prasad, who incidentally was the first president of the Indian Republic, wrote to Raman inviting him to be his personal guest in the Rashtrapathi Bhavan when he (Raman) came to Delhi for the award ceremony. You know what Raman did? He wrote a polite letter saying sorry he could not go. Now isn't that an insult to the President? Not at all. Raman had a definite reason for his inability to attend the investiture ceremony. He explained to the President that he was guiding a PhD student and that the thesis was positively due by the last day of January. The student was valiantly trying to wrap it all up, and Raman felt he *had* to be by the side of his student, see that the thesis was finished, sign the thesis as the Guide, and then have it submitted.

Just think about it! Here was a man who gave up the pomp of a glittering ceremony associated with the highest honour of the land because he felt that his duty required him to be by the side of his student. That is character. In the ultimate analysis, it is character that truly builds a nation. Raman valued the future of the country more than he valued awards given to him, and symbolic of that, he stayed with his student. That is true sacrifice.

Everyone cannot be a Raman but we certainly can give something back to our society for the many benefits it has silently conferred on us. It is no use being merely a great scientist. One must also be a good human being, concerned especially about the less fortunate and deeply conscious of all that one receives as well as all that one owes. Recall what Weisskopf said and think about it.

It has been nice being with you all along. Good luck, and may God bless you!

Index

W

X

Z